The Economics of the Tropical Timber Trade

Edward B Barbier, Joanne C Burgess, Joshua Bishop and Bruce Aylward

EARTHSCAN
Earthscan Publications Ltd, London

To Arthur Morrell, with many thanks and much appreciation.

First published in 1994 by
Earthscan Publications Limited
120 Pentonville Road, London N1 9JN

Copyright © Barbier, Burgess, Bishop and Aylward, 1994

All rights reserved. No part of this book may be reproduced in any form without written permission from the publisher and author.

A catalogue record for this book is available from the British Library

ISBN: 1 85383 219 7 pb/1 85383 226 X hb

Typeset by Books Unlimited (Nottm), Pleasley, NG19 7QZ
Printed and bound by Clays Ltd, St Ives plc

Earthscan Publications Limited is an editorially independent subsidiary of Kogan Page Limited, and publishes in association with the International Institute for Environment and Development and the World Wide Fund for Nature.

CONTENTS

List of Acronyms and Abbreviations   v
Acknowledgements   vii
Preface   ix

1 Introduction   1

2 Tropical Forest Resources and International Trade   5
The status of forest resources 5; Changes in forest resources 7; Production and trade 8; Future trends in supply and demand 16; Conclusions and policy implications 19

3 The Timber Trade and Tropical Deforestation   21
Is deforestation an economic problem? 21; Trade as an incentive for forest management 22; Environmental impacts of timber production 24; The links between trade and deforestation 29; Conclusions 32

4 The Market for Tropical Timber Products   34
Industrialized countries 34; Tropical developing countries 39; Consumer prices, substitution and demand 42; Conclusions 53

5 Tropical Forest Policies and the Environment   58

6 Domestic Market and Policy   61
The effect on forest management 61; Concessions and property rights 65; Forest pricing and its administration 67; Macro-economic policy 72; Conclusion 76

7 Trade Interventions in Exporting Countries   82
Restrictions by exporting countries 82; The impact of export restrictions 86; Trade liberalization by exporting countries 89; Domestic environmental policy 90; Conclusion 95

8 Trade Interventions in Importing Countries   97
Restrictions by importing countries 97; The impact of import restrictions 104; Trade liberalization by importing countries 106;

Impacts of environmental regulation by importing countries 108; Conclusions 111

9　Measures for the Sustainable use of Forests　　**114**
Identifying and assessing options 114; The feasibility of unilateral interventions 118; The feasibility of multilateral interventions 125: Conclusions 126

10　Assessment of Trade Policy Options　　**128**
Do nothing 128; Measures to alter the pattern of trade 129; Raising revenues for sustainable forest management 140; Certification 151; Conclusions 155

Epilogue　　**161**

References　　*163*
Index　　*177*

LIST OF ACRONYMS AND ABBREVIATIONS

ADC	apparent Domestic Consumption
AFP	all forest products
ATC	government-run river transport, Congo
CAR	Central African Republic
CBA	cost-benefit analysis
CFA	Communité Francophone Afrique franc (currency)
CFCO	Government-run railway, Congo
CGTM	CINTRAFOR Global Trade Model
CIB	Congolaise Industrielle des Bois
cif	cost, insurance and freight
CINTRAFOR	Centre for International Trade in Forest Products
CITES	Conference on the International Trade in Endangered Species
ECE	European Commission for Europe
EEC	European Economic Community
ERP	effective rate of protection
FAO	Food and Agricultural Organization of the UN
fob	free on board
FOE	Friends of the Earth
FPMO	Forest Protection and Management Obligation
FRIM	Forestry Research Institute of Malaysia
FSC	Forest Stewardship Council
GATT	General Agreement on Tariffs and Trade
GDP	Gross Domestic Product
GEF	Global Environmental Facility
GHK	Gewerkschaft Holz und Kunststoff, Germany
GNP	gross national product
GoC	Government of Congo
GoI	Government of Indonesia
GSP	Generalized System of Preferences
HDH	Central Federation of German Timber and Plastics Processing Industries
HIID	Harvard Institute for International Development
IIASA	International Institute for Applied Systems Analysis
IIED	International Institute for Environment and Development
IEEP	Institute for European Environmental Policy

INER	International Environmental Reporter
ITTA	International Tropical Timber Agreement
ITTC	International Tropical Timber Council
ITTO	International Tropical Timber Organization
JAWIC	Japan Wood-Products Information and Research Centre
LEEC	London Environmental Economics Centre
LKS	Lesser-Known species
MFN	Most-Favoured Nation
mn	million
mt	metric tonne
NEI	Netherlands Economic Institute
NGO	Non-Governmental Organization
NIC	Newly Industrialized Country
NPV	net present value
NRP	Nominal Rate of Protection
OECD	Organization for Economic Cooperation and Development
OFI	Oxford Forestry Institute
PNG	Papua New Guinea
R&D	research and development
Rp	Indonesian rupiah (currency)
SBH	Stichting Bos en Hout, The Netherlands
TEDB	Timber Export and Development Board of Ghana
TEV	Total Economic Value
TFAP	Tropical Forest Action Plan
TQ	Tariff/quota
TRADA	Timber Research and Development Association
TSC	Tropical Science Centre (Costa Rica)
TSM	Timber Supply Model
UCBT	Union pour le Commerce des Bois Tropicaux, European Union
UK	United Kingdom
UN	United Nations
UNCED	United Nations Conference on Environment and Development
UNEP	United Nations Environment Programme
UNU	United Nations University
USA	United States (of America)
USAID	United States Agency for International Development
USDA	United States Department of Agriculture
VAT	Value Added Tax
VDH	German Timber Importers Federation
WIDER	World Intstitute for Development Economics Research
WKS	Well-Known Species
WRI	World Resources Institute
WSC	Wood-products Stockpile Corporation
WTO	World Trade Organization
WWF	World Wide Fund for Nature

ACKNOWLEDGEMENTS

We wish to thank the International Tropical Timber Organization (ITTO) for their support, in particular the technical and administrative assistance provided by Steve Johnson and Jimmy Aggrey-Orleans of the ITTO Secretariat. However, while acknowledging this support, we wish to point out that the views are those of the authors alone and do not necessarily reflect the official position of ITTO, of any of its member countries, or of the ITTO Secretariat.

This book is the outcome of an Activity (PCM(XI)/4) supported by the ITTO. The original report was prepared with the assistance of Camille Bann, to whom we are grateful for this and other work that she conducted during her time at the London Environmental Economics Centre (LEEC). There were also numerous consultants who assisted us, many of whom contributed work that was eventually included in the technical annexes that accompanied the report to ITTO. Where possible, their contributions have been either reproduced in the LEEC Discussion Paper series or published elsewhere. Reference to this work is made throughout the book. We have benefited greatly from working with our consultants, and this book certainly reflects their important contributions and influence. Our consultants were Nancy Bockstael (University of Maryland), Jim Bourke (Food and Agriculture Organization of the United Nations), David Brooks (US Department of Agriculture Forest Service), Bruce Lippke (Centre for International Trade of Forest Products (CINTRAFOR)), Arthur Morrell, Juan Perez-Garcia (CINTRAFOR), Duncan Poore, Michael Rauscher (University of Kiel) and Ivar Strand (University of Maryland).

We also wish to acknowledge the contribution of those individuals who participated in the Workshops that were held at LEEC over 1992 to discuss the progress of the ITTO Activity and the policy options for encouraging the sustainable management of tropical forests: M Adigbli, J Aggrey-Orleans, R Barreto, S Bass, R Cooper, G Elliot, B Howard, P Landymore, S Meys, H Obdeyn, C Sargent and H Stoll. In addition, we would like to thank the many organizations and individuals with whom we held policy discussions and who assisted us in our search for data, in particular the Timber Export and Development Board of Ghana (TEDB), the Forestry Department of Ghana and Congolaise Industrielle des Bois (CIB), Congo. We would like to thank all other numerous individuals and organizations who have provided comments, references and support in the preparation of this report. Not least, we express our warm gratitude to Vinetta Holness and Jacqueline Saunders of LEEC who assisted us throughout the research and production of this book, and to Richard Sandbrook, Executive Director of the International Institute for Environment and Development (IIED), for promoting the work of LEEC and IIED generally and for being himself specifically.

Finally, a special thanks to Hans Obdeyn of the The Netherlands Ministry of Economic Affairs, who believed sufficiently in the merits of our research from the beginning to assist us in funding through ITTO auspices, and to Ron Kemp and the UK Delegation to ITTO for proposing this to ITTO in the first instance. However, it is our greatest regret that our most important supporter, from the early days of promoting the idea of this study in the UK Tropical Forest Forum to the final stages of policy discussion of our recommendations at the 15th Session of the International Tropical Timber Council, is no longer with us to witness the publication of this book. Sadly, Arthur Morrell died during the 15th Session, and it is with great fondness and appreciation that we dedicate this book to his memory.

PREFACE

The book is based on a report prepared for the International Tropical Timber Organization (ITTO) Activity (PCM(XI)/4) *The Economic Linkages Between the International Trade in Tropical Timber and the Sustainable Management of Tropical Forests*. The report was prepared and submitted by the London Environmental Economics Centre (LEEC) for review at the 14th Session of the International Tropical Timber Council (ITTC) in Kuala Lumpar, 11–17 May 1993. The main purpose of this Activity was to assess how the trade influences the incentives for sustainable management of tropical production forests, and to evaluate alternative trade policy options available to ITTO and its member countries to implement the 'Year 2000 Strategy' as outlined by Council Decision (X).[1] This Decision, undertaken at the 10th ITTC Session in Quito, Ecuador in May 1991, called on all ITTO member countries to adopt effective instruments for directly altering the incentives and disincentives for sustainable management of tropical forests, and in particular to encourage all trade in timber to be from sustainably managed sources by the year 2000. Activity PCM(XI)/4 and the subsequent report to ITTO by LEEC was seen to be an important step in implementing the Year 2000 Strategy.

To make the book more accessible as well manageable in terms of length, the technical annexes of the original report have been excluded and the book is based solely on the main Activity report to ITTO. Where possible, references are made to the relevant LEEC Discussion Papers and other published material for many of the details that were in the original technical annexes of the report.

This book examines the extent to which the tropical timber trade and policies, compared to other factors, affect tropical deforestation and timber-related forest degradation. It also analyses the potential role that interventions in the international tropical timber trade may have in promoting efficient and sustainable resource use in the forestry sector. We hope that it is presented in a sufficiently clear and non-technical style to be of interest to any person wanting to learn more about the economic linkages between the timber trade and tropical deforestation. As we have discovered, there are many misconceptions about this linkage.

For example, despite popular perceptions to the contrary, evidence on the link-

---

1 (Specifically, as stated in Decision 3(X), the 'Year 2000 Target' was established to encourage 'ITTO members to progress towards achieving sustainable management of tropical forests and trade in tropical forest timber from sustainably managed sources by the year 2000'.)

ages between the tropical deforestation, timber production and the timber trade suggests that the trade is not a major source of tropical deforestation. Not only is the conversion of forests to other uses such as agriculture a more significant factor, but an increasing proportion of tropical timber harvested in producer countries is for domestic consumption and does not enter international trade. For example, only 17 per cent of total non-coniferous tropical roundwood production is for industrial purposes. Of this, only 31 per cent is exported in round or product form. Therefore, only 6 per cent of total tropical non-coniferous roundwood production enters the international trade.

Thus the volume of tropical timber production that actually enters the trade is small and declining. As tropical timber resources are depleted and domestic consumption of timber products increases in producer countries, log prices are expected to rise and exports fall. However, increased exploitation of temperate timber resources, and possibly substitution by other wood and non-wood products, will keep the prices of tropical hardwood products down. Profit margins of tropical log processors will be squeezed, and export markets difficult to expand. Thus the main link between deforestation and the tropical timber trade may be the impact of increasing scarcity of tropical timber resources, and thus tropical hardwood logs, on production and exports.

Nevertheless, there is genuine cause for concern over the excessive exploitation and rapid depletion of tropical forests in many regions, including the indirect impacts of 'unsustainable' harvesting practices on the loss of non-timber forest values and the incentives to convert forest land to other uses (eg agriculture, livestock ranching). However, a more significant impact may be the role of 'unsustainable' timber production in opening up forest areas to subsequent agricultural encroachment and deforestation. These effects do not necessarily in themselves support strong statements about the relationship between timber production for the trade and tropical deforestation. As noted above, only a small proportion of tropical logs produced enters the trade directly, and in the future declining log exports are expected to be only partially offset by increased product exports.

On the other hand, the timber trade can lead to greater net returns for forestry investments and sustainable management of production forests, making this option more attractive than alternative uses of forest land such as agriculture. Unfortunately, in many producer countries the widespread prevalence of market and policy failures have distorted the incentives for sustainable management. Failures in concession and pricing systems have produced counterproductive incentives that lead to the 'mining' of production forests. Domestic market and policy failures have also had a major influence on the conversion of forest land to agriculture and other uses.

Thus the key factor in reducing timber-related tropical deforestation is ensuring proper economic *incentives* for efficient and sustainable management of tropical production forests. Appropriate forest management policies and regulations within producer countries ought to provide these incentives so that the *long run* income-generating potential of harvesting timber is maximized, and any significant external environmental costs associated with timber harvesting are 'internalized'. The starting point for any strategy to promote a sustainably managed timber trade, such as ITTO's Target 2000, is therefore to tackle the problem at its source by urging producer countries to improve forest sector policies.

Current trade policy distortions in producer and consumer countries have, if anything, exacerbated the problems created by poor forestry policy and regulations in tropical forest countries. Although log export restrictions in producer countries have stimulated growth and employment in domestic processing, they have led to serious problems of processing over-capacity and inefficiency. To the extent that this is the case, log prices are artificially depressed and recovery rates fall, thus increasing pressure on timber resources. Although import tariffs on tropical forest products are generally low and declining in major developed consumer markets, non-tariff barriers may be significant and increasing. Restrictions on imports depress the global demand for tropical timber products and can feed back to reduce stumpage values in producer countries, thus discouraging the incentives for more efficient processing and better forest management. Moreover, producer countries will continue to argue that they need to compensate their domestic processing through subsidies and export restrictions.

In short, restrictions in trade are not helping to reduce timber-related deforestation in developing countries. In contrast, by adding value to forestry operations, the trade in tropical timber products *could* act as an incentive to sustainable production forest management – provided that the appropriate domestic forest management policies and regulations are also implemented by producer countries. Unfortunately, many proposed trade policy interventions to 'save the tropical forests' – such as bans, taxes and quantitative restrictions – may actually work to *restrict* the trade in tropical timber products. Such interventions may reduce rather than increase the incentives for sustainable timber management – and may actually increase overall tropical deforestation. A summary of timber trade policy options, and their advantages and disadvantages in promoting sustainable tropical forest management is provided on page 156.

However, the book does suggest that there is a role for additional tropical timber trade policies in fostering *trade-related incentives* for sustainable management. Trade policies will be the most effective if:

- they are employed in conjunction with and complement improved domestic policies and regulations for sustainable forest management within producer countries
- they improve rather than restrict access to import markets for tropical timber products so as to ensure maximum value added for sustainably produced tropical timber exports
- they assist producer countries in obtaining the additional financial resources required to implement comprehensive national plans for sustainable management of tropical production forests.

In the book, we provide much detail on the economic linkages between the timber trade and tropical deforestation that reinforce the above conclusions. We do not expect everyone to agree with us. We certainly believe that there is scope for much more work in this area. Some of this work might produce equally compelling evidence and arguments that support different conclusions from the ones we have derived. What matters to us is not the outcome of future research but that there is sufficient interest in the economics of the tropical timber trade and its links with tropical deforestation to stimulate further research in the first place. However, our greatest hope is that the international community will find the 'political will' to follow up on the evidence and analysis presented in this book, and perhaps in future research, by enacting appropriate policy responses – before it is too late for much of the world's tropical forests.

Edward B Barbier
Joanne C Burgess
Joshua T Bishop
Bruce A Aylward

York and London
19 July 1994

1

INTRODUCTION

Concerns about tropical deforestation have led to an increased focus on the role of the timber trade in promoting forest depletion and degradation. Recent reports suggest a marked increased in tropical deforestation in the 1980s, with the overall rate doubling from 0.6 per cent in 1980 to 1.2 per cent in 1990 (Dembner 1991). Although the deforestation rate varies across regions, with an estimated annual rate for Latin America of only 0.9 per cent compared with 1.7 per cent for Africa and 1.4 per cent for Asia, the loss of forest area has increased significantly in all tropical regions.

Despite concern over the state of tropical deforestation and its implications for global welfare, several recent studies have indicated that the tropical timber trade is not the major *direct* cause of the problem compared to conversion of forests for agriculture.[1] Nevertheless, it is clear that current levels of timber extraction in tropical forests – both open and closed – exceed the rate of reforestation (WRI 1992). Less than 1 million hectares, out of an estimated total global area of 828 million hectares of productive tropical forest in 1985, was under sustained-yield management for timber production (Poore et al 1989). Moreover, timber extraction has a major *indirect* role in promoting tropical deforestation by opening up previously unexploited forest, which then allows other economic uses of the forests such as agricultural conversion to take place (Amelung and Diehl 1992a). For example, in many African producer countries, around half of the area that is initially logged is subsequently deforested, while there is little, if any, deforestation of previously unlogged forested land (Barbier 1994).

Some of the environmental values lost through timber exploitation and depletion, such as watershed protection, non-timber forest products, recreational values, may affect only populations in the countries producing the timber. Concerned domestic policymakers in tropical forest countries should therefore determine whether the benefits of incorporating these environmental values into decisions affecting timber exploitation balance the costs of reduced timber production and trade, as well as the costs of implementing such policies. What economists call the socially 'optimal' level of timber exploitation and trade – ie, what is more popularly referred to as 'sustainable' timber management – is the level of exploitation that 'internalizes', or incorporates, the additional domestic environmental costs of

---

1   See, for example, Amelung and Diehl (1992); Barbier (1994); Barbier et al (1991); Vincent and Binkley (1991); Hyde, Newman and Sedjo (1991); Repetto (1990); and Repetto and Gillis (1988).

logging the forests, wherever it is feasible to do so. Designing policies to control excessive forest degradation is clearly complex and requires careful attention to harvesting incentives. As recent reviews suggest, many domestic policies in tropical timber producing countries do not even begin to approximate the appropriate incentives required to achieve a socially optimal level of timber harvesting. More often than not, pricing, investment and institutional policies for forestry actually work to *create* the conditions for short-term harvesting by private concessionaires, and in some instances, even *subsidize* private harvesting at inefficient levels.[2] Over the long term, incentive distortions that understate stumpage values and fail to reflect increasing scarcity as old growth forests are depleted can undermine the transition of the forestry sector from dependence on old growth to secondary forests and the coordination of processing capacity with timber stocks (Binkley and Vincent 1991).

Increasingly the world's tropical forests, including their remaining timber reserves, are also considered to provide important 'global' values, such as a major 'store' of carbon and as a depository of a large share of the world's biological diversity (Barbier et al 1994; Pearce 1990; Reid and Miller 1989). Similarly, even some 'regional' environmental functions of tropical forests, such as protection of major watersheds, may have transboundary 'spillover' effects into more than one country. But precisely because such transboundary and global environmental benefits accrue to individuals outside of the countries exploiting forests for timber, it is unlikely that such countries will have the incentive to incur the additional costs of incorporating the more 'global' environmental values in forest management decisions. On the other hand, individuals in the rest of the world who are concerned about the loss of global environmental benefits through tropical deforestation are increasingly demanding that all tropical forests – including those with timber production potential – be managed so that such global benefits can be 'sustained'. Not surprisingly, sanctions and other interventions in the timber trade are one means by which other countries may seek to coerce timber producing countries into reducing forest exploitation and the subsequent loss of global environmental values. In addition, trade measures are increasingly being explored as part of multilateral negotiations and agreements to control excessive forest depletion, to encourage 'sustainable' timber management and to raise compensatory financing for timber producing countries that lose substantial revenues and incur additional costs in changing their forest policy.

However well intentioned they may be, both domestic and international environmental regulations and policies that attempt to 'correct' forest management decisions may have high economic, and even 'second order' environmental, costs associated with them (Barbier 1994). There is increasing concern that the potential trade impacts of environmental policies that affect forestry and forest-based industries may increase inefficiencies and reduce international competitiveness. Moreover, the trade impacts of domestic environmental regulations may affect industries in other countries and lead to substantial distortions in the international

---

2   For example, see Barbier (1994); Barbier et al (1991); Vincent and Binkley (1991); Gillis (1990); Hyde, Newman and Sedjo (1991); Repetto (1990); and Repetto and Gillis (1988).

timber trade. The overall effect on the profitability and efficiency of forest industries may be to encourage forest management practices that are far from 'sustainable'. Careful analysis of both domestic and international environmental policies affecting forest sector production and trade is therefore necessary to determine what the full economic and environmental effects of such policies might be.

The starting point for such an analysis is a better understanding of the *economic linkages* between the trade in tropical timber products and deforestation in tropical regions. Only through exploring such linkages is it possible to begin unravelling some of the important issues and concerns over the role of timber exploitation and trade in tropical deforestation, and to suggest appropriate policy measures for encouraging sustainable management of production forests. It is this theme that is the underlying rationale of the book.

As a matter of practicability, we concentrate our efforts primarily on the analysis of current trade conditions and policies rather than on forest management practices and the conditions to achieve sustainable management within the timber production sector. Hence, our main concern is with the linkages between the trade in tropical timber products, tropical deforestation and sustainable forest management.

However, as noted above, the latter term – 'sustainable' forest management – has been interpreted in different ways by various individuals concerned with the problem of tropical deforestation. We use the term here in a more narrow sense of sustainable production and environmental management of those forests currently or potentially exploitable for timber in tropical regions rather than in the broader sense of the 'sustainability' of the entire tropical forest area, or of all the potential 'values' generated by this area. Thus our working definition of sustainable forest management is the definition and criteria of sustainability adopted by the International Tropical Timber Organization (ITTO):

> Sustainable forest management is the process of managing permanent forest land to achieve one or more clearly specified objectives of management with regard to the production of a continuous flow of desired forest products and services without undue reduction in its inherent values and future productivity and without undue undesirable effects on the physical and social environment.[3]

Consequently, in this book our primary concern will be with management of the *production forests*, within the overall permanent forest estate, of producer countries. Similarly, the forest resource base of principal concern is the tropical timber resources of producer countries, ie their standing *hardwood* resources. Finally, the forest products, or *tropical timber products* of interest are the industrial wood products generated from these hardwood resources in tropical countries. These include logs, sawnwood, plywood (including veneers) and other wood products.

The book examines the extent to which the tropical timber trade and policies, compared to other factors, affect tropical deforestation and timber-related forest degradation. The book also analyses the potential role that interventions in the international tropical timber trade may have in promoting efficient and sustainable resource use in the forestry sector. The outline of the rest of the book is as follows.

---

3   From the ITTC Decision 6(XI), Quito, 8th Session, May, 1991. See also ITTO (1990).

Chapter 2 provides a brief overview of the historic, current and projected status of tropical forest resources and international timber trade flows. Chapter 3 examines in more detail the specific contribution of the tropical timber trade, compared to other causal factors, in tropical deforestation and the wider environmental impacts of tropical forest use. Chapter 4 provides an important insight into the market conditions for timber products in both producer countries and importing countries and a survey of end-users of tropical timber products in the UK. These three chapters raise a number of key policy issues concerning the environmental effects of the tropical timber trade, forestry policy and wider economic policies, which are an important background to the analysis that follows in the remainder of the book and is summarized in Chapter 5.

Chapter 6 discusses domestic market and policy failures, both within the forestry sector and economy-wide, that impact on tropical forest management. The current trade interventions in tropical timber exporting and importing countries are explored in Chapters 7 and 8 respectively, including an examination of the impacts of trade restrictions on tropical forests. Chapter 9 defines the international timber trade policy options, given the current state of forest management and trade interventions; provides a list of criteria for choosing between these options; and discusses the rationale for implementing trade policy options on a multilateral as opposed to a unilateral basis. Chapter 10 assesses each of the proposed trade policy options to determine which are more likely to provide trade-related incentives for sustainable management of tropical forests in producer countries. Finally, Chapter 11 provides an overall conclusion to the analysis and findings of the book, and looks to the prospects in the future of implementing sustainable timber management.

2
TROPICAL FOREST RESOURCES AND INTERNATIONAL TRADE

This chapter consists of a synthesis of existing data and studies to provide a statistical review of tropical forest resources and the international tropical timber trade. This brief overview provides a background for the rest of the book, especially the formulation and analysis of policy options for the international tropical timber trade to encourage sustainable forest management.[1]

THE STATUS OF FOREST RESOURCES

Table 2.1 depicts the status and changes of natural tropical forest resources across ecological zones, regions and selected countries (*Food and Agricultural Organization of the United Nations* (FAO) 1992; Schmidt 1990).[2] The total area of natural tropical forests was estimated to be 1,715 million hectares (mn ha) in 1990 – approximately 36 per cent of the total land area in the tropics. Of this, tropical rainforests (656 mn ha) and moist deciduous forests (626 mn ha) constitute the largest portions (38 per cent and 37 per cent respectively).

On a regional level, Africa accounts for 47 per cent of the total land area within the tropics, but contains only 35 per cent of the total tropical forest. Asia is the smallest region within the tropics (15 per cent of the total tropical zone area), but it contains an equal proportion of tropical forest (15 per cent of total tropical forest area). Latin America and the Caribbean make up the remaining 35 per cent of the tropical zone area, and contain the largest extent of tropical forest resources (50 per cent of total tropical forest area). Within the tropical zone, over half of the closed forest area is concentrated within three countries, namely Brazil (347 mn ha), Zaire (103 mn ha) and Indonesia (109 mn ha).

The extent of land under plantations in the tropics is relatively small compared to that under natural forests – in 1990 total plantation area was 43.9 mn ha, less than 2 per cent of the total tropical forest area.[3] Industrial plantations (mainly

---

1 A more comprehensive review and additional tables can be found in Annex A of the Technical Annex of the original LEEC report to ITTO, henceforth referred to as Barbier et al (1993).

2 The FAO (1992) tropical forest data is based on information from 87 tropical countries which contain over 97 per cent of the world's tropical forests.

Table 2.1 Tropical forest resources: status and changes (000 ha)

	Land area	Forest area 1990	Area deforested annually 1981-90	% of area deforested annually 1981-90
Main ecological regions[a]				
Lowlands				
Tropical rainforest	912,000	655,500	4,900	0.75
Moist deciduous forest	1,461,100	626,400	7,300	1.17
Dry deciduous forest	720,500	212,900	2,100	0.99
Very dry forest	547,700	39,500	200	0.51
Desert	523,800	2,500	100	4.00
Uplands				
Hill and montane forest	650,500	178,100	2,300	1.29
Total	**4,815,600**	**1,714,900**	**16,900**	**0.99**
Regions[a]				
Africa	2,243,300	600,100	5,000	0.83
Asia	896,600	274,900	3,600	1.31
Latin America and Caribbean	1,675,700	839,900	8,300	0.99
Total	**4,815,600**	**1,714,900**	**16,900**	**0.99**
Selected Countries[b]				
Latin America				
Brazil	845,651	347,000	3,200	0.92
Peru	128,000	73,000	300	0.41
Bolivia	108,439	55,500	60	0.11
Venezuela	88,205	42,000	150	0.36
Colombia	103,870	41,400	350	0.85
Guyana	19,685	19,300	3	0.02
Suriname	15,600	15,200	3	0.02
Ecuador	27,684	12,300	60	0.49
Africa				
Zaire	226,760	103,800	200	0.19
Congo	34,150	21,100	22	0.10
Gabon	25,767	20,300	15	0.07
Cameroon	46,540	17,100	80	0.47
Central African Republic	62,298	3,600	5	0.14
Equatorial Guinea	2,805	1,200	3	0.25
Asia				
Indonesia	181,157	108,600	1,315	1.21
Malaysia	32,855	18,400	255	1.39
Philippines	29,817	6,500	110	1.69

Sources [a]FAO (1992)
[b]Schmidt (1990)

3   The definition of plantation used here is taken from FAO (1992) which states that 'a plantation is a class of forest established artificially: (i) on land which previously did not carry forest; or (ii) involving the replacement of the previous crop by a new and essentially different species'.

eucalyptus, pine and teak) make up less than 35 per cent of total plantation area, the majority of which is located in the Asia and Pacific region (58 per cent), followed by Latin America (33 per cent), with the remaining 9 per cent in Africa (FAO 1992).

CHANGES IN FOREST RESOURCES

The *extent* of tropical deforestation reached 16.9 mn ha per annum (pa), at an annual deforestation *rate* of 0.9 per cent, throughout the 1980s.[4] Tropical deforestation was concentrated in Latin America (8.3 mn ha pa) and Africa (5 mn ha pa). Asia experienced the lowest extent of tropical deforestation (3.6 mn ha pa), but the highest rate of deforestation at 1.2 per cent. Among the ecological zones, 7.3 mn ha of forests were deforested each year in the moist forest zone, 4.9 mn ha pa in the tropical rainforest zone and approximately 2 mn ha pa in both the dry and hill/montane zones, with the moist deciduous forest zone and the hill/montane forest zone suffering the highest annual rate of deforestation at 1.1 per cent (see Table 2.1).

Of the tropical countries, Brazil (3.2 mn ha) and Indonesia (1.3 mn ha) incur the highest *extent* of annual forest loss (Table 2.1). However, the *rates* of tropical deforestation are highest in those countries that have high annual losses of tropical forests combined with relatively small forest resources, such as Ivory Coast (6.5 per cent), Nigeria (5 per cent), Costa Rica (4 per cent), Paraguay (4.7 per cent) (World Resources Institute (WRI) 1992). Due to the vast extent of their forest stocks, the annual rate of deforestation in the 'Big Three' tropical forest countries remains relatively low: in Brazil it is 0.9 per cent; in Indonesia 1.21 per cent; and, in Zaire 0.2 per cent.

Forests, of course, are not only destroyed but are regenerated naturally as secondary forests, regenerated by human intervention or renewed naturally by human management. Reafforestation refers to three separate processes:

- The restoration of previously existing forests (artificial regeneration)
- The artificial establishment of a new forest on previously forested ground (reforestation)
- The artificial establishment of forests in previously unforested areas (afforestation).

The area planted each year amounts to approximately 2.6 mn ha, which is dramatically outstripped by the level of annual deforestation (16.9 mn ha) in the tropics.

It is extremely difficult to generalize about the implications of the reduction in tropical forest resources for the timber trade. In some regions, the decrease in the

---

4 Measurements of the extent of 'forest change' are heavily influenced by the definition that is employed. In this chapter we look only at forest change according to the strict FAO definition of deforestation, that is 'the change of land use or depletion of crown cover to less than 10 per cent' (FAO 1988). However, Chapter 3 discusses in more detail the implications of using other definitions of forest change, such as degradation and modification, when analysing the causes of tropical deforestation.

natural tropical forest resources constrain the availability of timber for the trade. In other regions, natural tropical forest resources are still plentiful, but economic factors such as extraction and transportation costs constrain the supply of timber for the trade. Whilst an increase in industrial plantations may offset the reduction in supply of timber from natural forests in the future, this depends on a rapid increase in the level of investment in reafforestation over the next few years. The type of species planted, and the degree of substitutability between plantation timber (usually softwood species) and timber from natural forests (mainly hardwoods), will also influence the extent to which plantations can offset the loss of natural tropical forests. An additional concern is that the 'wider' economic benefits of tropical forests (eg non-timber forest products, watershed protection, biodiversity value, carbon store, climate regulation) are being irreversibly lost through tropical deforestation. These themes will be discussed in greater detail throughout the rest of this book.

PRODUCTION AND TRADE

The production of, and trade in, timber products on a *global* level has been expanding throughout the last few decades.[5] The volume of industrial roundwood produced throughout the world has grown steadily to reach 1,654 mn cubic metres (m^3) in 1990, of which around 7 per cent enters the international trade as industrial roundwood (Table 2.2). Given the focus of this study on the tropical timber trade, it is important to note that only around 30 per cent (in volume terms) of the total global production of industrial forest products is derived from non-coniferous forests which are mainly located in tropical regions.

Trade

Of the total world non-coniferous tropical roundwood production (1,400 mn m^3 in 1990), only 17 per cent is used for industrial purposes with the remaining being consumed as fuelwood and for other non-industrial uses (Table 2.3a).[6] Out of the total volume of industrial timber produced (275 mn m^3), approximately 31 per cent (86 mn m^3) is exported in round or product form by tropical countries. Therefore, only a small percentage (6 per cent) of total tropical non-coniferous round-

---

[5] The FAO segregates timber products into five general categories: roundwood, sawnwood and sleepers, wood-based panels, wood pulp, paper and paperboard. The sub-category 'roundwood' is composed of both 'industrial roundwood' (ie saw and veneer logs, pulpwood and particles) and 'non-industrial roundwood' (ie fuelwood and charcoal). All further processed products (such as furniture, window frames and doors) are referred to as 'higher processed products'.

[6] Tropical timber is defined here as non-coniferous timber coming from developing countries lying predominantly between the Tropic of Cancer and the Tropic of Capricorn. However, due to data constraints some of the following tables are based on coniferous and non-coniferous timber data. Where this switch occurs, it is clearly indicated in the text and tables.

Table 2.2a Volume of world production and trade in forest products (mn m³)[a]

	1961	1970	1980	1990
Fuelwood and charcoal				
Production	1,041	1,186	1,480	1,796
Imports	4	3	3	4
Imports as a % of production	0.3	0.2	0.2	0.2
Industrial roundwood				
Production	1,018	1,278	1,452	1,654
Imports	38	93	118	124
Imports as a % of production	3.8	7.3	8.1	7.5
Sawnwood and sleepers				
Production	346	415	451	486
Imports	41	56	78	95
Imports as a % of production	11.8	13.6	17.2	19.5
Wood-based panels				
Production	26	70	101	125
Imports	3	10	16	29
Imports as a % of production	12.0	14.4	15.5	23.4
Wood pulp				
Production	62	102	126	155
Imports	10	17	21	25
Imports as a % of production	16.0	16.3	16.4	16.3
Paper and paperboards[1]				
Production	77	126	170	238
Imports	13	23	34	55
Imports as a % of production	16.5	18.0	19.9	22.9

Note [a]Wood pulp and paper and paperboards in mn metric tonnes (mt).
Source FAO (1992)

Table 2.2b Values of world trade in forest products (US$ mn, current)

	1961	1970	1980	1990
All forest products (AFP)				
Imports	6,778	14,170	62,377	123,360
Fuelwood and charcoal				
Imports	31	29	122	187
as % of AFP	0.5	0.2	0.2	0.2
Industrial roundwood				
Imports	908	2,693	12,316	12,523
as % of AFP	13.4	19.0	19.7	10.2
Sawnwood and sleepers				
Imports	1,841	3,020	13,952	34,395
as % of AFP	27.2	21.3	22.4	27.9
Wood-based panels				
Imports	437	1,184	5,236	10,391
as % of AFP	6.4	8.4	8.4	8.4
Wood pulp				
Imports	1,316	2,650	9,777	17,341
as % of AFP	19.4	18.7	15.7	14.1
Paper and paperboards				
Imports	2,244	4,567	20,845	48,252
as % of AFP	33.1	32.2	33.4	39.1

Source FAO (1992)

Table 2.3a The volume of world and tropical timber production and trade, 1990

	Production	Exports	Tropical production as a % of world production	Tropical exports as a % of world production
Total roundwood (mn m³)				
World	3,450.4	120.5		
Tropical	1397	30	40.5	24.9
Industrial roundwood (mn m³)				
World	1,654.2	118.2		
Tropical	275	29.1	16.6	24.6
Saw and veneer logs (mn m³)				
World	979.8	66.8		
Tropical	149	27	15.2	40.4
Sawnwood and sleepers (mn m³)				
World	485.9	88.9		
Tropical	59	9	12.1	10.1
Wood-based panels (mn m³)				
World	124.9	31.2		
Tropical	19.4	13	15.5	41.7
Wood pulp (mn tonnes)				
World	154.4	25		
Tropical	7.2	0.7	4.7	2.8
Paper and paperboard (mn tonnes)				
World	238.2	55.2		
Tropical	15.5	2.2	6.5	4.0

Note Tropical countries are all developing countries excluding China, Chile, Argentina, Turkey and South Korea

Table 2.3b The value of world and tropical timber production and trade, 1990

	Exports (US$ bn)	Tropical as a % of world exports
Saw and veneer logs		
World	6.57	
Tropical	2.28	34.7
Sawnwood		
World	16.99	
Tropical	2.15	12.7
Wood-based panels		
World	10.14	
Tropical	4.15	40.9
Wood pulp		
World	15.82	
Tropical	0.86	5.4
Paper and paperboard		
World	45.27	
Tropical	1.5	3.3
Other		
World	2.68	
Tropical	0.16	6.0
Total		
World	97.47	
Tropical	11.1	11.4

Note Tropical countries are all developing countries excluding China, Chile, Argentina, Turkey and South Korea
Source Bourke (1992)

wood production enters the international trade (Bourke 1992). Exports from tropical developing countries account for a modest share of the total value of world exports (US$ 11 billion (bn) out of US$ 97.5 bn, ie 11 per cent).

Table 2.3b lists (in value terms) the trade balance for tropical countries that currently export forest products.[7] Tropical Asia and Oceania is the major export-

Table 2.4 (cont)

	Imports	Exports	Net Exports
Tropical Africa			
Cameroon	35,412	99,833	64,421
Cent Afr Rep	468	29,994	29,526
Congo	4,500	106,087	101,587
Côte d'Ivoire	27,200	236,147	208,947
Eq Guinea	0	18,700	18,700
Gabon	3,655	136,774	133,119
Ghana	5,129	76,526	71,397
Guinea	1,056	800	−256
Guinea-Bissau	310	350	40
Kenya	23,594	4,054	−19,540
Liberia	1,942	78,264	76,322
Madagascar	8,546	534	−8,012
Malawi	8,085	1,993	−6,065
Mozambique	950	923	−27
Nigeria	33,083	1,680	−31,403
Sierra Leone	1,028	146	−882
Tanzania	15,700	1,539	−14,161
Zaire	3,666	17,032	13,366
Zimbabwe	5,765	4,169	−1,596
Tropical C and S America			
Belize	3,253	2,445	−808
Costa Rica	40,020	21,895	−18,125
Cuba	193,411	1,847	−191,564
El Salvador	21,800	2,725	−19,075
Guatemala	69,410	18,326	−51,084
Honduras	137,921	31,061	−106,860
Mexico	403,605	13,884	−389,721
Nicaragua	10,566	2,569	−7,997
Panama	76,979	3,988	−72,991
Trinidad and Tobago	54,396	458	−53,938
Bolivia	4,060	22,160	18,100
Brazil	299,402	1,750,981	1,451,579
Colombia	104,056	20,060	−83,996
Ecuador	157,834	24,373	−133,461
French Guiana	1,087	2,169	1,082
Guyana	2,356	2,694	338
Paraguay	13,055	24,971	11,916
Peru	104,914	2,558	−102,356
Suriname	9,671	840	8,831

7 Note that in the table the term 'forest products' includes industrial and non-industrial wood products and does not distinguish between coniferous and non-coniferous timber.

Table 2.4 (cont)

	Imports	Exports	Net Exports
Tropical Asia and Oceania			
Brunei Darus	6,775	30	−6,745
Cambodia	100	94	−6
Hong Kong	1,752,273	705,535	−1,046,738
India	290,967	16,337	−274,630
Indonesia	330,157	3,069,199	2,739,042
Laos	200	10,251	10,051
Malaysia	483,372	3,040,884	2,557,512
Myanmar	4,721	148,084	143,363
Philippines	173,662	123,119	−50,543
Singpore	747,548	663,302	−84,246
Sri Lanka	28,771	600	−28,171
Thailand	1,002,371	101,551	−900,820
Yemen	10,499	29	−10,470
Fiji	7,804	22,775	14,971
Papua New Guinea	5,504	115,500	109,996
Solomon Islands	767	17,240	16,473
Vanuatu	202	1,900	1,698

Note *Tropical countries taken here to be countries with the majority of their land mass lying between the tropics. The term 'forest products' includes industrial and non-industrial wood products. Only those countries that exported forest products in 1990 are included in this table.
Source FAO (1992)

ing region, followed by Africa, with Central and South America playing a minor role in the international forest products trade. The major exporting countries are Indonesia, Malaysia and Brazil, Gabon and Congo who each exported over US$ 1000 mn in forest products in 1990.

Whilst the production of tropical roundwood has been steadily increasing over the past few decades, timber consumption has been growing at an even faster rate due to population and income growth. The rapid expansion of timber demand in many tropical countries has led to increased domestic consumption, reduced timber exports, and increased timber imports. As a result, many tropical timber producing countries are becoming net timber importers. The forest products trade generates positive net foreign exchange earnings in only 10 of the 18 African exporting countries, 5 of the 19 Central and Southern American exporting countries, and 8 of the 17 exporting countries from Asia and Oceania (Table 2.4).

Estimates (in volume terms) of the proportion of timber products that enter world trade from tropical countries compared to their apparent domestic consumption (ADC) of these products are presented in Table 2.5a.[8] As noted above, tropical Asia and Oceania is clearly the most important region in the forest products trade – it accounts for half of the industrial roundwood, sawnwood and wood-based panels produced by tropical countries and its exports represent over 85 per cent of total exports of these products from tropical countries. Tropical Central and South America produce 36 per cent of forest products from tropical countries, but consume most of this domestically. Tropical Africa produces a minor share of

---

8 In Table 2.5 the data refers to coniferous and non-coniferous timber products.

Table 2.5a Production and trade in timber products by tropical countries, 1990 (000 m³)[a]

	Production	Exports	Imports	ADC[b]
All Tropical Countries				
Industrial roundwood	257,587	28,705	4,318	233,200
Sawnwood	72,584	8,719	4,841	68,706
Wood-based panels	18,483	12,818	1,891	7,556
Tropical Africa				
Industrial roundwood	41,687	3,959	40	37,768
Sawnwood	6,598	814	235	6,019
Wood-based panels	1,108	243	66	931
Tropical C and S America				
Industrial roundwood	95,697	172	95	95,620
Sawnwood	26,641	1,032	1,557	27,166
Wood-based panels	4,289	833	296	3,752
Tropical Asia and Oceania				
Industrial roundwood	120,203	24,574	4,183	99,812
Sawnwood	39,345	6,873	3,049	35,521
Wood-based panels	13,086	11,742	1,825	3,169

Table 2.5b Exports of timber products as a percentage of production in tropical countries (%)

	1961	1970	1980	1990
All Tropical Countries				
Industrial roundwood	15.6	27.1	18.2	11.1
Sawnwood	15.3	17.3	16.2	12.0
Wood-based panels	34.0	33.0	32.5	69.4
Tropical Africa				
Industrial roundwood	23.8	23.3	16.3	9.5
Sawnwood	32.1	28.5	12.7	12.3
Wood-based panels	34.0	33.0	28.6	21.9
Tropical C and S America				
Industrial roundwood	1.6	1.0	0.2	0.2
Sawnwood	14.7	13.6	6.9	3.9
Wood-based panels	34.0	33.0	15.3	19.4
Tropical Asia and Oceania				
Industrial roundwood	22.2	44.2	33.4	20.4
Sawnwood	11.9	18.5	24.5	17.5
Wood-based panels	40.5	39.9	49.4	89.7

Notes: [a]Tropical countries are taken here to be countries with the majority of their land mass lying between the tropics. This table includes data on coniferous and non-coniferous timber products.
[b]ADC is apparent domestic consumption, or production + imports − exports.
Source: FAO (1992)

forest products from tropical countries (14 per cent) and exports roughly 10 per cent of its domestic production. Whilst Africa accounts for around 16 per cent of the ADC of roundwood in all tropical countries, its proportionate ADC of sawnwood (9 per cent) and wood-based panels (12 per cent) is slightly lower.

Table 2.5b shows exports of timber products as a percentage of domestic production in tropical regions between 1961 and 1990. On average in 1990, tropical regions exported 11 per cent of industrial roundwood, 12 per cent of sawnwood, and 69 per cent of wood-based panels, with the rest consumed domestically. The proportion of industrial roundwood and sawnwood exported relative to production has decreased since the 1970s, while in contrast the proportion of wood-based panels exported relative to production has doubled. However, this pattern is dominated by the Asian bloc – in particular the effect of the dramatic increase in the percentage of wood-based panels exported relative to production rising from 40 per cent in the 1960–70s to 90 per cent in 1990 – and is not applicable to all regions. For example in Africa, while exports of industrial roundwood and sawnwood as a percentage of production have fallen over time, exports of wood-based panels have also fallen from 34 per cent in 1961 to 22 per cent in 1990. This is mainly due to the increased domestic consumption of processed wood in Africa. Similarly, in Central and South America, exports of all forest products compared to production have decreased since the 1960s.

Real Price Trends

The real forest product price index has been fairly stable over the 1969–88 period, although revealing a gradual downward trend since the mid-1970s (Figure 2.1).[9] Within the broad forest products category, the real price of tropical logs has followed a rising trend since the early 1970s; prices briefly declined between 1979 and 1985, only to pick up again thereafter (Figure 2.2). The real price of tropical sawnwood has followed a similar trend, although with larger fluctuations and a steeper decline in the mid-1980s (Figure 2.3). The real price of other industrial timber products, including wood-based panels, pulp, and paper and paperboard, have sustained this rise throughout the 1980s. The real price increases may reflect increasing product scarcity due to declining forest inventories and increasing demand for tropical timber products, whilst the recent downturn in some real prices may reflect the depressed state of the global economy during the late 1970s and early 1980s.

There are regional differences in the export price of tropical timber products. For example, the average unit export price of non-coniferous logs from Africa (price US$ 136/m^3 in 1988) is substantially higher than the average unit export price of non-coniferous logs from Asia (current price US$ 78/m^3 in 1988) (FAO 1990). Regional price differentials are explained, to some extent, by the composi-

---

9   The real prices are derived from current export unit values and deflated with an index of export unit values of manufactured goods from the *UN Monthly Bulletin of Statistics*. The forest products price index includes both coniferous and non-coniferous forest products traded internationally.

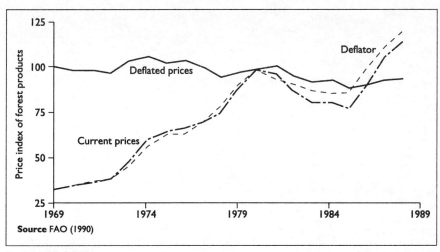

Figure 2.1 Forest products price index (US$/m³, 1980 = 100)

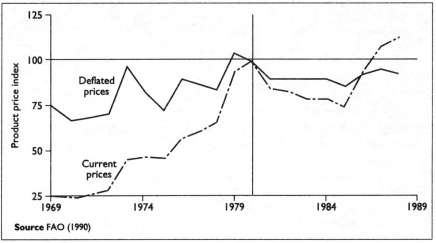

Figure 2.2 Tropical logs price index (US$/m³, 1980 = 100)

tion of species exported. Chapter 4 will look in more detail at the market conditions for timber products.

The Value of the Trade to Tropical Regions

As noted above, only a small proportion of tropical timber harvested is used for industrial purposes (17 per cent) and an even smaller proportion of total roundwood production enters the international trade (6 per cent). The industrial forest sector and timber trade is, however, important to tropical countries for a number of reasons:

- First, the industrial forest sector makes a direct contribution to the economy of producer countries – for example, in 1989 the industrial forest sector

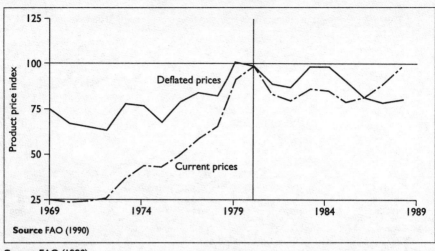

Source FAO (1990)

Figure 2.3 Tropical sawnwood price (US$/m³, 1980 = 100)

accounted for 3–6 per cent of total gross domestic product (GDP) in Malaysia and Indonesia, and 2–3 per cent of GDP in Ivory Coast, Gabon, Ghana, Brazil and Costa Rica. However, in most other tropical timber producing countries the industrial forest sector represented less than 2 per cent of GDP in 1989 (World Bank 1992).

■ Second, wood-related industries are an important source of employment generation in the manufacturing sector of many developing countries, although the level of employment in wood-related industries is often extremely modest compared to the size of economically active populations in developing countries.

■ Third, forest product exports are a valuable source of foreign exchange for a few countries. For example, forest product exports account for over 10 per cent of the total value of exports from Central African Republic (CAR), Ghana, Indonesia, Malaysia and Papua New Guinea. However, across the tropics as a whole, exports of forest products do not generate substantial foreign exchange earnings compared to total export earnings (Table 2.6).

■ Finally, there are numerous wider economic and social benefits associated with timber production and trade that are also important for tropical developing countries, such as rural infrastructure development and the provision of other social amenities.

FUTURE TRENDS IN SUPPLY AND DEMAND

A number of studies have attempted to forecast future trends in supply and demand in the forest sector on a global and a regional level.[10] A comprehensive review of

---

10 See Sedjo and Lyon 1990; Kallio et al 1987; FAO 1990; Cardellichio et al 1989; ECE/FAO 1986, 1989 and 1990; and USDA Forest Service 1990.

Table 2.6 Forest products trade compared to total trade in tropical countries, 1990

	Forest Products Imports (US$ 000)	Forest Products Exports (US$ 000)	Total Imports (US$ 000000)	Total Exports (US$ 000000)	Forest Products Imports as a % of Total Imports (%)	Forest Products Exports as a % of Total Imports (%)
Tropical Africa						
Cameroon	35,412	99,833	1,300	1,200	2.7	8.3
CAR	468	29,994	170	130	0.3	23.1
Congo	4,500	106,087	570	1,130	0.8	9.4
Côte d'Ivoire	27,200	236,147	2,100	2,600	1.3	9.1
Gabon	3,655	136,774	760	2,471	0.5	5.5
Ghana	5,129	76,526	1,199	739	0.4	10.4
Kenya	23,594	4,054	2,124	1,033	1.1	0.4
Madagascar	8,546	534	480	335	1.8	0.2
Malawi	8,058	1,993	576	412	1.4	0.5
Nigeria	33,083	1,680	5,688	13,671	0.6	0.0
Sierra Leone	1,028	146	146	138	0.7	0.1
Tanzania	15,700	1,539	935	300	1.7	0.5
Zaire	3,666	17,032	888	999	0.4	1.7
Tropical C and S America						
Costa Rica						
El Salvador	40,020	21,895	2,026	1,457	2.0	1.5
Guatemala	21,800	2,725	1,200	550	1.8	0.5
Honduras	69,410	18,326	1,626	1,211	4.3	1.5
Mexico	137,921	31,061	1,028	916	13.4	3.4
Nicaragua	403,605	13,884	28,063	26,714	1.4	0.1
Panama	10,566	2,569	750	379	1.4	0.7
T&T	76,979	3,988	1,539	321	5.0	1.2
Bolivia	54,396	458	1,262	2,080	4.3	0.0
Brazil	4,060	22,160	716	923	0.6	2.4
Colombia	299,402	1,750,981	22,459	31,243	1.3	5.6
Ecuador	104,056	20,060	5,590	6,766	1.9	0.3
Paraguay	157,834	24,373	1,862	2,714	8.5	0.9
Peru	13,055	24,971	1,113	959	1.2	2.6
	104,914	2,558	3,230	3,277	3.2	0.1
Tropical Asia and Oceania						
Hong Kong						
India	1,752,273	705,535	82,495	29,002	2.1	2.4
Indonesia	290,967	16,337	23,692	17,967	1.2	0.1
Malaysia	330,157	3,069,199	21,837	25,553	1.5	12.0
Philippines	483,372	3,040,884	29,251	29,409	1.7	10.3
Singapore	173,662	123,119	13,080	8,681	1.3	1.4
Sri Lanka	747,548	663,302	60,647	52,627	1.2	1.3
Thailand	28,771	600	2,689	1,984	1.1	0.0
PNG	1,002,371	101,551	33,129	23,002	3.0	0.4
	5,504	115,500	1,288	1,140	0.4	10.1

Note T&T = Trinidad and Tobago; PNG = Papua New Guinea
Source FAO (1992); and World Bank (1992[b])

many of these studies is provided by Arnold (1991), also supported by Vincent (1991). The major trends on long term supply and demand of timber are summarized below:

- Demand for industrial wood is projected to grow at a rate of 15 to 40 per cent over the next 15 years. Developing countries will take an increasing share of this growth, although developed countries will continue to dominate the market.
- Changes in technology and preferences will favour the growth in consumption of reconstituted products as opposed to that of solid products such as sawnwood.
- Hardwood consumption will grow faster than softwood consumption.
- Tropical resources will decline, but expanding temperate resources will be more than sufficient to lead to stable real prices for wood products.
- Projected growth in demand will be met by plantations; 'unexpected' increased in demand could be met by additional resources in central Canada, northern Europe and eastern Russia.
- The shift to consumption of planted and second growth forests from old growth stands will continue. The focus will shift from the Pacific Northwest and the tropics to forests in the southern and northeastern US and newly planted southern temperate resources.
- Japan and western Europe could become increasingly self-sufficient, while parts of the developing world may need to increase their imports.

In his synopsis, Arnold (1991) suggests the tropics are expected to be increasingly less important in the overall trade picture. As temperate countries look more to temperate resources to meet their needs and domestic consumption in tropical producer countries grows, the international trade in tropical timber is likely to decline. Whilst the total *volume* of the tropical timber trade may decline, producer countries may export a higher proportion of value-added wood products, and thus any decline in the *value* of the trade may be less significant. In addition, an increase in South–South trade – particularly in sawnwood – will act as a counteracting force (Arnold 1991). Asia will continue to be the dominant producer and exporter of tropical timber, although with proportionately less log exports and more value-added timber products. As a result of Asian producers moving into downstream markets, Africa and Latin America may become relatively more important sources of log and sawnwood exports respectively.

Many of these conclusions are confirmed by more recent projections carried out for this study (Perez-Garcia and Lippke 1993). New projections of supply, demand and trade in hardwood timber products were made using a model of the global forest sector, developed by the Centre for International Trade in Forest Products (CINTRAFOR).[11] Baseline short term projections to the year 2000 reveal a number of significant trends:

---

11 For full details on the results of projections and policy simulations based on the latest version of the CINTRAFOR Global Trade Model (CGTM), using data up to 1990, see Perez-Garcia and Lippke (1993), or Annex K in Barbier et al (1993). See Chapters 7–9 for more discussion of the policy results of the CGTM simulations.

- Decreasing commercial inventory of tropical timber is beginning to constrain harvests significantly, particularly in Malaysia. There is no comparable shortage of temperate hardwoods, which supply large consuming country markets in the US, Europe and other non-Asian markets. Even with increased saw log production in Indonesia and Brazil, two tropical hardwood suppliers with a large inventory, the tropical hardwood share of all hardwoods will begin to decline.
- The combination of expected strong economic growth in tropical timber producing countries, more modest demand growth in other consuming countries, and declining supply of tropical hardwood logs will produce a substantial shift away from export to domestic markets by the major tropical hardwood suppliers. Declining log exports will be offset only partially by increased product exports.
- Declining tropical hardwood inventory will lead to steadily rising saw log prices, reaching levels 60 to 80 per cent above 1990 levels by the year 2000, in real terms. Most of this price increase is anticipated to occur in SE Asia. Prices in countries with adequate temperate hardwood sources will remain more stable.
- While product prices will increase with rising log prices, the availability of other supply sources in the more developed consuming countries, in conjunction with lower demand growth, will constrain product price increases which in turn will squeeze profits for processors of tropical timber. The trend in the developed consuming countries is therefore for a reduction in log imports and the processing of imported tropical logs, and increased product imports with more substitution away from tropical products.

Long term projections to 2040 show that the commodity in short supply continues to be tropical hardwood logs, not processing capacity. With the available tropical hardwood inventory in several countries declining rapidly by the year 2000, either harvest levels will be reduced quickly to more sustainable levels or they will drop even more abruptly just a few years later with the depletion of the inventory. Long term projections also suggest that tropical saw log prices will continue their upward trend but product price increases will not keep pace. While consumer countries are forced to accept slightly higher prices and do change their consumption patterns with those high prices, tropical log prices are unlikely to flatten out and decline significantly unless increased investment in sustainable forest management can support increased harvest levels. Higher prices for tropical logs should stimulate such investment in the long run.

CONCLUSIONS AND POLICY IMPLICATIONS

There are a few key issues with important policy implications which arise from this review of forest resources and the international trade in timber products:

- The rate of tropical deforestation is significant (0.9 per cent for 1980–90) and has increased in recent years. However, the decline in the stocks of tropical

timber resources is being offset by expanding temperate resources, especially from plantations.
- Only 17 per cent of tropical non-coniferous wood is used for industrial purposes. Out of the total volume of industrial timber (275 mn m^3), approximately 31 per cent is exported in round or product form by tropical countries. Therefore, only a modest percentage (6 per cent) of total tropical non-coniferous roundwood production enters the international trade.
- Tropical Asia is the most important region in terms of volume of timber produced (over 50 per cent of tropical production) and exported (over 85 per cent of tropical exports). Asia is also taking the lead in processing timber products, and higher valued timber products account for an increasing proportion of its exports.
- Exports from tropical developing countries account for a small share of the total value of world timber exports (US$ 11 bn out of US$ 97.5 bn, ie 11 per cent). However, for a few countries (eg CAR, Ghana, Indonesia, Malaysia and Papua New Guinea) timber exports are an important source of foreign exchange earnings (ie over 10 per cent of total export earnings).
- Domestic consumption of timber by tropical countries will continue to increase due to population and income growth. Many tropical timber producers are becoming net timber importers. The level of South–South trade is expected to expand over the next few years.
- Demand for industrial timber on a global level is expected to increase up until the year 2000. Developing countries will account for an increasing share of this growth, although developed countries will continue to dominate the global timber market.
- Declining tropical timber inventory combined with demand growth in producer countries will push up log prices and decrease log exports. Expanding temperate resources will prevent commensurate increases in the prices of hardwood products. Processors of tropical logs will see their profits squeezed, leading to declining imports of tropical logs into consumer countries. This is only partially offset by increased product imports from tropical countries.

Given the small and declining volume of tropical timber production that actually enters the international trade, there appears to be little direct leverage to use the trade as a means of encouraging sustainable tropical forest management. On the other hand, while the value of the tropical timber trade is not particularly significant for most exporting countries, it is an important source of foreign exchange earnings and wider socioeconomic benefits for a few key producer countries. The loss of benefits associated with the timber trade may undermine incentives for tropical countries to manage and conserve their forest resources. This theme will be examined in more detail in Chapter 3.

3

The Timber Trade and Tropical Deforestation

This chapter discusses the links between the timber trade and tropical deforestation.[1] An important question to be asked at the outset is whether tropical deforestation can be considered as an economic problem. We then focus on the role of the timber trade as an incentive for sustainable tropical forest management, which involves looking at alternative tropical forest land use options and the economics of land allocation. The chapter draws together the literature on the direct and indirect links between timber production and tropical deforestation. Existing statistical analyses of the relationship between timber production, the trade and forest clearance in the tropics are reviewed, and a new analysis presented. Finally, the policy implications for the international tropical timber trade as a means for encouraging sustainable forest management are summarized.

IS DEFORESTATION AN ECONOMIC PROBLEM?[2]

To answer this question requires addressing two additional issues:

- Are tropical forests an efficient form of holding on to 'wealth'?
- Are the opportunity costs of tropical deforestation greater than the benefits gained?

Although apparently different, the questions are inherently related and represent two aspects of the same fundamental economic problem concerning tropical deforestation.

Tropical forests can be seen as a form of 'natural' capital. That is, they have the potential to contribute to the long run economic productivity and welfare of tropical forest countries. Too often, decisions concerning tropical forest depletion and conversion are taken without considering the opportunity cost of these decisions. In short, we do not know whether it is worth 'holding on' to the tropical forest as an economic asset because we do not know or bother to take into account the potential economic benefits that the natural forest resource yields. As a result,

---

1   A more comprehensive analysis of these links is provided by Burgess (1993) and (1994).

2   The first two sections of this chapter are drawn from Barbier (1993).

decisions will always be biased towards converting or depleting the tropical forest because the underlying assumption is that the benefits of maintaining the forest are negligible. If this is the case, as much evidence suggests, then current levels of tropical deforestation may be 'excessive' from an economic standpoint.

TRADE AS AN INCENTIVE FOR FOREST MANAGEMENT

Land Use Options

If tropical deforestation is an economic problem, it is because important values are lost, some irreversibly, when closed forests are opened up, degraded or cleared. Each choice or *land use option* for the forest – to leave it standing in its natural state, or to exploit it selectively, eg for timber or non-timber forest products, or to clear-cut it entirely so the land can be converted to another use, such as agriculture – has implications in terms of values gained and lost. The decision as to what land use option to pursue for a given tropical forest area, and ultimately whether current rates of deforestation are 'excessive', can only be made if these gains and losses are properly analysed and evaluated. This requires that *all the values* that are gained and lost with each land use option are carefully considered.

Although it is the responsibility of governments to ensure that the total economic value of forest land lost through conversion and exploitation is accounted for in development decisions, most evidence suggests that public policies are often at the core of the tropical deforestation problem.[3] Too often, the pricing and economic policies of countries with tropical forests distort the costs of deforestation:

- The 'prices' determined for tropical timber products or the products derived from converted forest land *do not incorporate the lost economic values* in terms of foregone timber rentals, foregone minor forest products and other direct uses (eg tourism), disrupted forest protection and other ecological functions, and the loss of biological diversity, including any option or existence values.
- Even the direct costs of harvesting and converting tropical forests *are often subsidized and/or distorted*, thus further encouraging forest conversion and degradation.

If proper economic valuation of forest losses takes place, it can be shown that the cost of forest conversion and degradation is often high. For example, in Indonesia, the foregone cost in terms of timber rentals from converting primary and secondary forest land is in the order of US$ 625–750 mn pa With logging damage and fire accounting for additional costs of US$ 70 mn, this would represent losses of around US$ 800 mn pa. The inclusion of foregone non-timber forest products,

---

3   For recent comparative reviews of how public policies affect tropical deforestation, see Barbier (1994); Barbier, Burgess and Markandya (1991), Vincent and Binkley (1991); Hyde, Newman and Sedjo (1991); Repetto (1990a); and Repetto and Gillis (1988).

such as essential oils, honey, wildlife products, resins, bamboos, fruits and nuts, would raise this cost to US$1 bn pa The value of exports of non-timber forest products rose from US$ 17 mn in 1973 to US$ 154 mn in 1985, comprising 12 per cent of export earnings. Exports of rattan alone were US$ 80 mn in 1985. However, the commercial potential of many non-timber products may be currently under-exploited. The foregone future value of these products may therefore be much higher than their current value indicates. Moreover, many of the important economic uses of non-timber products, eg the use of 'wild' forest foods to supplement food security and meet subsistence needs, are non-marketed and thus are not often properly accounted for. Consequently, both the foregone commercial and non-marketed benefits of non-timber products must be incorporated into any assessment of the costs of deforestation (Pearce et al 1990, Chapter 5).

The total cost of the depreciation of the forest stock would include not just the cost of conversion but also the cost of timber extraction and forest degradation. One study estimated this total cost for Indonesia to be around US$ 3.1 bn in 1982, or approximately 4 per cent of GDP (Repetto and Gillis 1988). However, this estimate must be considered a lower bound, as it does not include the value of the loss of forest protection functions (eg watershed protection, micro-climatic maintenance) and of biodiversity. The latter may particularly be important in terms of *option and existence values* – i.e values reflecting a willingness to pay to see species conserved for future use or for their intrinsic worth – which could translate into future payments that the rest of the world might make to Indonesia to conserve its forest lands. Similar arguments apply to the value of the forest as a *carbon store* (Pearce 1990).

The Economics of Land Use Decision Making

The decision over which tropical forest land use option to pursue depends on the relative returns from the different uses. The following simplified example is used to illustrate the economics of land use decision making. Consider the choice between clearing the land for agricultural production and sustainably managing the forest for timber production, some of which enters the international timber trade. If the discounted net returns of sustainable timber production (NPV^T) exceed the net returns from converting the forest for agricultural production (NPV^A), then it will be in the direct economic interest of the individual, or society, to sustainably manage the land for timber production.

$$NPV^T > NPV^A$$

That is, so long as sustainable forest management for timber production leads to economic rent greater than that derived from forest clearance for agricultural production, then there exists an incentive to conserve the forest for timber production.[4] Increasing the potential value of timber production through the export of timber products will increase the overall net returns to tropical forest management,

---

4 Economic rent is the net return to a unit of land. This value is the residual, or surplus, remaining after the costs of all other factors of production are netted out.

thus making it more attractive compared to converting forest land to agriculture. Moreover, if timber production for export is done on a sustainable basis, then sustainable management of tropical forests has the chance of being an economically attractive alternative to competing forest land uses over the medium to long term.

The above example is extremely simplified and does not reflect the full complexities of decision making in practice (such as a range of alternative land use options, other political objectives, conflicts between individual and society decisions, valuation of costs and benefits). However, it does illustrate an extremely important point – *that in order for sustainable timber management to be a viable forest land use option, it must yield net returns to developing countries that are greater than those derived from competing uses.*[5] In subsequent chapters, we demonstrate that trade related incentives are an important means of ensuring appropriate returns from sustainable timber management of tropical forests.

ENVIRONMENTAL IMPACTS OF TIMBER PRODUCTION

Although it is possible to manage the extraction of timber from tropical forests on a sustainable basis with minimum environmental damage, this practice does not appear to be widespread. In a comprehensive survey of management practices throughout tropical forest countries, Poore et al (1989) indicated that less than 1 mn ha, out of an estimated total area of 828 mn ha of productive tropical forest remaining in 1985, was demonstrably under sustained-yield management for timber production. This does not necessarily mean that 99 per cent of productive forests are being degraded in the tropics through timber extraction, but is more an indication of the lack of sustainable forest management occurring in the tropics. It is the environmental problems related to the unsustainable and poorly managed extraction of timber that are of concern. These may occur:

- *directly* through the removal of trees and other damage incurred to surrounding forest during timber extraction; and
- *indirectly* through opening up and improving access to the forests which then impacts on other socioeconomic factors which may degrade the environment.

Direct Environmental Impacts

Timber production and the international trade in tropical timber products is reasonably well documented. However, it is extremely difficult to extrapolate from

---

5   In arguing against a timber trade boycott, Vincent (1990) emphasizes that the forest must be *used* if it is to be saved; that is, 'in the tropics as elsewhere, forests must not compete with other land uses to remain wooded. A boycott would reduce demand and depress forest product prices. This would reduce net returns for forestry investments and make sustained timber management, a prerequisite for stablization of forest areas in the tropics, less feasible'.

such information to the direct impact of the international trade on the tropical forests for a number of reasons, including:

- uncertainty over the magnitude of forest degradation, conversion and regeneration in many tropical countries;
- little information on the impact of forest management regimes and their implementation on tropical forests;
- substantial methodological problems in assessing the wider environmental implications of timber extraction practices; and
- difficulty in determining the amount of timber extracted for domestic use as opposed to that entering international markets.

In addition to depleting wood resources, timber extraction can incur *external environmental costs* by degrading other tropical forest resources and functions which are of value to individuals other than timber operators. As discussed above, these 'external' effects may include the loss of other consumptive uses (eg harvesting and hunting other forest resources and recreational uses), of ecological functions (eg watershed protection, carbon storage and micro-climatic role) and of other non-consumptive values (eg ecotourism, genetic resource and existence values) of the forest. Much of the criticism of tropical deforestation stems from scientists' claims that closed tropical forests are estimated to hold between 50–90 per cent of the world's biodiversity (Reid and Miller 1989). Evidence of the importance of the sustainable uses of biodiversity for subsistence forest products, pharmaceutical and crop breeding research, ecotourism, local watershed protection, micro-climatic functioning and the maintenance of other important environmental functions as well as overall ecosystem integrity means that extensive timber extraction in the tropics can impose significant external costs (Barbier, Burgess and Folke 1994).

The extent of these external environment impacts from timber extraction depends largely on the type, and success, of forest management practices. Poore et al. (1989) note that successful forest management depends on certain conditions being met, including the long term security of operation, operational control, a suitable financial environment and adequate information. Although there are a few cases of successful sustained-yield management of forests for timber production – eg in some regions in India, Malaysia and the Philippines – these tend to be the exception. Poorly designed and implemented management regimes for selective logging of natural forests are likely to have serious implications for the environment.

For example, on average only about 5 to 35 m^3 of merchantable wood are extracted per ha of tropical closed broadleaved forests (FAO/UNEP 1981). However, these small commercial volumes relative to total standing timber can lead to disproportionate damage to the forest due to careless use of equipment and inefficient logging practices. Sometimes at least half the remaining stock, including immature trees of potential commercial value and harvestable stocks of less desirable varieties, are damaged beyond recovery (Repetto 1990b). Clear-cutting the forest for the timber is likely to have even more significant environmental effects. However, measuring the wider environmental impacts and costs of forest management practices is complicated, and few empirical studies have been undertaken.

The direct impact of timber production on the environment may be offset to some extent by investments in *reafforestation*. However, even if investments to offset deforestation are channelled into plantation forests then only part of the full environmental costs of timber extraction may be compensated. That is, investments in plantations may counteract the decline in stocks of timber, but may not always compensate for the wider environmental costs of natural forest degradation. Even if the compensatory investment is in management of a natural forest area, some of the environmental benefits of the degraded natural forest may still have been lost irreversibly, such as biodiversity or non-timber forest products.

Indirect Environmental Impacts

In addition to the direct impacts of timber extraction on the environment, tropical timber production can influence environmental degradation *indirectly*. This indirect impact may occur through the opening up and improvement of access to the forests, which may then interact with other socioeconomic factors encouraging activities that degrade the environment. However, due to the intricately interconnected relationship of the various causes of deforestation, it is extremely difficult to identify how much of the tropical deforestation process is due to timber production alone.

In many developing countries where there are still areas of previously unexploited forest, and there exist no formal property rights for this land, the private returns to timber production may encourage open access exploitation at the forest frontier and rapid forest conversion. Timber extraction usually involves extensive road-building which benefits other activities, such as agriculture and hunting, by improving access to the forest and reducing costs of transporting produce to market. In Northern Brazilian Amazon, the total road network (paved and unpaved) increased from 6,357 to 28,431 kilometres (km) over 1975–88 period. Although the road expansion programme cannot be specifically attributed to timber extraction in this case, a simple correlation between road density and the rate of deforestation demonstrates that as road density increases, the rate of deforestation increases in larger proportions (Reis and Marguilis 1991).

Timber extraction is often the first step towards opening up the tropical forest and clearing the land for agricultural production. What is more, in many developing countries, property law establishes deforestation as a prerequisite of formal claim over the land for those settling in forested areas (Mahar 1989, Pearce et al 1990). Table 3.1 suggests around half of the area logged in African countries is subsequently deforested, whilst there is little, if any, deforestation of previously unlogged forest land. The environmental impact of forest conversion to agriculture has grown as logging has progressively opened up more remote, hilly and ecologically vulnerable areas.

Table 3.1 Timber harvesting and deforestation in African ITTO producer countries, 1981–5 (000 ha)

Country	Area logged[a]	Area logged deforested[b]	Unlogged area deforested[c]
Cameroon	272.0	75.0	3.0
Congo	57.0	20.0	1.5
Côte d'Ivoire	330.0	290.0	[d]
Gabon	150.0	15.0	[d]
Ghana	na	22.0	na
Liberia	104.0	44.0	[d]
Total	>913	466.5	>4.5

Notes [a]Total average area selectively logged per annum
[b]Estimated area of selectively logged land that is subsequently deforested
[c]Unlogged area deforested per annum
[d]Unknown but small
Source Barbier, Rietbergen and Pearce (1994)

A study by Amelung and Diehl (1992) looked at the causes of tropical forest *clearance, degradation and modification* (Table 3.2). This study clearly shows that the use of different definitions of 'tropical forest change' can lead to very different conclusions about the causes of deforestation. The more sensitive the definition to forest change, the more likely the effect of the forestry sector is to be highlighted. For example, using the FAO definition of *deforestation* it appears that the agricultural sector accounted for the largest share (over 80 per cent) of total deforestation in all tropical forest countries during the 1980s, and the direct impact of forestry activities on deforestation is minimal (ie 2 per cent of total deforestation).[6] However, when alternative, more sensitive, measures of forest degradation and forest modification are used, the role of the forestry sector becomes much more significant. For example, in Indonesia the forestry sector accounts for over 40 per cent to total biomass reduction (*forest degradation*) and for all major tropical countries it accounts for 10 per cent. The forestry sector is responsible for the vast majority of *forest modification* (71 per cent across all major tropical countries) by converting previously unexploited forests into productive closed forests or other forms of land use. This analysis also indicates a process of deforestation, whereby timber extraction is largely responsible for the first stages of opening up previously unexploited forest, which then enables other economic uses of the forest resources, such as agricultural cultivation, which lead to further forest degradation and then deforestation.

---

6 As noted in the previous chapter, the FAO employs a classification of deforestation to be a change of land use or a depletion of crown cover to less than 10 per cent. Changes *within* the closed forest classification are termed forest degradation (or biomass reduction). However, there does not appear to be any clear distinction made between forest degradation and biomass reduction, although the two are not strictly synonymous.

Table 3.2a Sources of deforestation in tropical countries, 1981–88[a]

	Brazil	Indonesia[b]	Cameroon	All major tropical countries
Forestry	2[d]	9	0	2 (10)[e]
Agriculture	89	80	100	(83)f
Shifting cultivators[c]	13 (23)	59 (67)	92 (95)	na (47)
Permanent agriculture	76	21	8	36
Pastures	40	0	0	17
Permanent crops	4	2	5	3
Arable land	32	19	3	16
Mining including related industries	<3	<0.3	0	na
Hydroelectricity production	4[b]	0	0	2[b]
Residual[g]	2	11	0	(13)[h]

Notes
[a] Percentage shares in deforestation refer to averages for the respective period.
[b] Data refer to 1980–90.
[c] Figures in parentheses show the results of the FAO for 1980. These results include also market oriented farmers who produce cash and export crops and engage only partly in shifting cultivation.
[d] Deforestation due to logging is due to charcoal production.
[e] The figure in parentheses refers to an alternative estimation. The calculation includes only Indonesia and Brazil, since these countries account for the largest share in clear-cutting by the forestry sector.
[f] This percentage rate is based on the assumption that the percentage share calculated for shifting cultivators can be taken as an average for 1981–8.
[g] This includes other industries, housing, infrastructure services and fire loss.
[h] The residual has been calculated from the data in this column which includes data from different periods.

Table 3.2b Sectoral share in forest degradation and modification, 1981–5[a]

Sector	Percentage share in biomass reduction (degradation)				Percentage share in forest modification			
	Brazil	Indonesia	Cameroon	Total[b]	Brazil	Indonesia	Cameroon	Total[b]
Forestry	6	44	10	10	(100)[c]	(100)[c]	98	71
Agriculture[d]	85	49	90	76	0	0	2	26
Others[d]	9	7	0	13	0	0	0	4

Notes
[a] For the definition of modification and biomass reduction (degradation) see source.
[b] Total refers to all major rainforest countries.
[c] Following FAO statistics, deforestation in virgin forests is 0, since clearing by other sectors concentrates on disturbed forests. Even though some clearing occurs in virgin forests, there is reason to assume that the bulk of deforestation is due to forests that have been logged over prior to the clearing of the respective areas.
[d] These figures reflect averages for 1981–8 or, in the case of Indonesia, 1980–90.

Source Amelung and Diehl (1992)

THE LINKS BETWEEN THE TIMBER TRADE AND DEFORESTATION

A number of studies have attempted to assess the relative importance of various economic activities, including timber extraction, in causing *tropical deforestation*. For example, Binswanger (1989) and Mahar (1989) highlight the role of subsidies and tax breaks, particularly for cattle ranching, in encouraging land clearing in the Brazilian Amazon. More recent analysis by Schneider et al (1990) and Reis and Marguilis (1991) emphasize the role of agricultural rents, population pressures and roadbuilding in encouraging small-scale frontier settlement in this region. The study by Schneider et al (1990) also identifies the importance of logging – log production from the Amazon region increased from 4.5 mn m^3 in 1975 to over 24.5 mn m^3 in 1987 – in forest exploitation, primarily through opening access to previously inaccessible lands. Commercial timber extraction has been encouraged by both a range of public policies, increasing domestic demand, and a strong international demand for tropical hardwoods. However, most of the work is extremely tentative and constrained to qualitative analysis because of the problems outlined earlier.

A few studies have used statistical analysis to explore the linkages among factors thought to cause forest land clearance. There are a number of important caveats that need to be borne in mind when reviewing the statistical studies of factors thought to contribute to deforestation. Firstly, all the analyses suffer from the usual problem of reliability and accuracy of data. Data limitations make it difficult to distinguish between production forests and conversion forests across the tropics. This may lead to misleading estimates of the relationship between log production and changes in forest area. If log production is mainly from conversion forests, then timber extracted is essentially a precursor or byproduct of agricultural conversion which is the principle factor in the resulting deforestation. In addition, due to the level of aggregation required to undertake such studies, the analyses are not usually sensitive to the different types of forests and different patterns of wood use. Finally, although many of the studies reviewed are relevant to the relationship between timber production and tropical deforestation, most were not designed explicitly to explore this relationship. Nevertheless, statistical approaches can provide interesting insights to the existent and relative importance of relationships among factors thought to cause deforestation. A brief overview of these studies is provided below.

- ■ A comprehensive cross-sectional analysis of the causes of deforestation in 39 developing countries in Africa, Latin America and Asia supports the hypothesis that forest loss between 1968 and 1978 increases with increasing 'wood use' (Allen and Barnes 1985). The measure of wood use is constructed from the sum of wood fuel consumption and wood exports – reflecting the demand for fuelwood as well as wood exports. The authors justify use of this composite variable on the grounds that the sum of the two factors are more closely related to deforestation than either wood fuel production and wood exports alone. It is therefore difficult to isolate the links between wood exports and deforestation from this study.

- A series of cross-sectional (72 tropical countries) statistical tests of various factors thought to influence deforestation was carried out by Palo, Mery and Salmi (1987). Taking total forest coverage in 1980 as the dependent variable, they tested the correlation between this and a variety of independent variables, including forest product exports per forest area and relative industrial roundwood production. Neither of these two independent variables was found to be statistically significant.
- Capistrano (1990) and Capistrano and Kiker (1990) examine the influence of macro-economic factors on the depletion of tropical forests in 45 tropical developing countries. The regression analysis was run over 4 periods – 1967–71, 1972–75, 1976–80, and 1981–5 – but was not over the entire 1967–85 period. The authors take changes in timber production area as a proxy for total deforestation. Although they argue there is a close correlation between average area of closed broadleaved forest and industrially logged forest, there are many countries where industrial logging is not a significant source of deforestation. Thus their analytical results are more relevant to deforestation of tropical timber production forests than to overall tropical deforestation. The export value of tropical wood is highly significant in the first period (1967–71) and explains more than 85 per cent of the variation in the area of productive forest logged during this period. However, it is not significant in any later periods.
- A regression analysis of deforestation in Indonesia using pooled cross-sectional and time series data was undertaken for this book (Barbier et al 1994; Barbier et al 1994).[7] This analysis shows that log production is associated with a reduction of forested area in Indonesia's forested provinces. The recursive deforestation model is linked to a simultaneous equation system which determines the demand and supply in the logging, sawnwood and plywood sectors of Indonesia. This is used to trace through the impacts of timber trade policies on the wood products markets, timber production and thus deforestation. However, data limitations prevent distinguishing between permanent production and conversion forest. Close to 50 per cent of Indonesia's total production forest consists of conversion forest and there is anecdotal evidence to suggest that log production mainly came from the conversion forest. Therefore, the deforestation equation may be reflecting the impact of agricultural conversion, rather than the initial logging operations, on changes in forested area and thus overestimating the impact of log production on deforestation.
- Building on the lessons learned in previous studies, a regression analysis to examine the relationship between timber production and forest clearance throughout the tropics has been developed for this book.[8] The analysis covers 53 tropical countries and looks at the relationship between a range of variables (ie timber production, agricultural yield, population density, income growth

---

7   Also contained in Annex I of Barbier et al (1993).

8   Further details and the results of this study can be found in Burgess (1993 and 1994) and in Annex B of Barbier et al (1993).

Table 3.3 The linkages between timber production and tropical deforestation

Dependent Variable	Five year change in closed forest area (logarithm forest area (000 ha) 1985 – logarithm forest area (000 ha) 1980)
Explanatory Variables	**Estimated Coefficient (t statistic)**
Constant	−0.1009 (−4.188)
X1 (logarithm of closed forest area as a percentage of total area 1980 (log of forest area 1000 ha (*100)/total area 1000 ha)	0.01253 (1.609)
X2 (population density 1980 (total population/total land area (ha))	−0.0474 (−2.695)
X3 (industrial roundwood production per capita 1980 (m^3/total population))	−0.0849 (−2.3029)
X4 (real GNP per capita 1980 (US$ GNP/total population))	0.000195 (1.8098)
X5 (agricultural yield 1980 (cereal production 1000 mt/cereal production area 1000 ha))	0.02301 (1.1298)
X6 (dummy Latin America)	−0.06809 (−2.4086)

Estimated elasticities
X1 = 0.0125
X2 = −0.0285
X3 = −0.0186
X4 = 0.1870
X5 = 0.0339
X6 = −0.0216
R^2 0.268
F Statistic: 2.8089

Source Burgess (1993) and (1994)

and tropical forest stock) and forest clearance. The statistical model (presented in Table 3.3) is fairly robust given the complexity of the problem being analysed, the aggregate level of the analysis, the cross-sectional nature of the model and data limitations. The model supports the hypothesis that industrial roundwood production is positively associated with forest clearance in the tropics for 1980–85 period – ie increasing levels of industrial roundwood production per capita leads to higher rates of forest loss.[9] A 1 per cent increase in roundwood production (m^3 of industrial non-coniferous roundwood produced per capita in 1980) is anticipated to increase the level of forest area cleared by 0.019 per cent. [10]

The statistical analyses reviewed above provide only limited evidence of the linkages between tropical timber production, trade and deforestation. This suggests that we need to be cautious when making statements about the role of the trade in tropical deforestation. Further research is required at a regional and country level to provide more conclusive results for policy makers. Other factors pertaining to the timber trade also suggest caution is required when developing appropriate policy interventions to encourage sustainable forest management. For example, Chapter 4 considers market conditions and the nature of demand for tropical timber and its implications for trade policy.

CONCLUSIONS

This chapter has discussed the linkages between the tropical timber trade and sustainable forest management. It is extremely difficult to identify and analyse the linkages for a number of reasons, in particular the complexity of the linkages, the lack of data on the status of forest resources, insufficient information on the returns to forest land under alternative uses, and the difficulty of isolating the impacts on the forest of timber harvested for the trade from total timber production. However, there are several important policy implications that can be drawn from this chapter:

- Current rates of tropical deforestation are thought to be excessive and tropical deforestation is considered to be an economic problem.
- In order for sustainable timber management to be a viable forest land use option it must yield net returns to developing countries that are greater than those derived from competing uses, such as conversion for frontier agriculture. Trade related incentives are an important means of ensuring appropriate returns to sustainable timber management of tropical forests and making it an economically attractive land use option.

---

9   The model was also run using the variable industrial non-coniferous roundwood production per total area in 1980. The coefficient on this alternative explanatory variable was also negative (–0.000167) and there was little change to the estimated coefficients of the other independent variables. However, the explanatory power of this alternative variable was slightly less significant (t statistic –1.68) and the robustness of the model in general was reduced through the variable's inclusion.

10 From Table 3.3, it is clear that population density is also positively and more strongly linked to deforestation – a 1 per cent increase in population density would increase forest clearance by 0.02 per cent. In contrast, economic development as represented by increasing income per capita and improvements in agricultural yields seem to reduce forest clearance significantly. Moreover, a country with a lower than average forest resource base in 1980 appears to have incurred higher levels of forest clearance, suggesting that many countries with relatively small forest areas to total land areas may be running down their remaining forest stocks at a fast rate.

- Although it is possible to manage the extraction of timber from tropical forests on a sustainable basis with minimum environmental damage this practice does not appear to be widespread. Timber production can degrade tropical forests directly through unsustainable forest management practices and indirectly through opening up the forests.
- The statistical analyses reviewed provide only limited evidence of the linkages between the tropical timber trade and tropical deforestation. This suggests that we need to be cautious when making statements about the role of the trade in tropical deforestation. However, as noted in Chapter 2, only a small proportion of the total volume of industrial roundwood produced is traded internationally. Therefore, the use of trade interventions to encourage improved forest management through reducing the volume of timber production is likely to be ineffective due to the limited role of the trade in overall timber production.

This chapter has highlighted a dichotomy in the linkage between the timber production for the trade and tropical deforestation. On the one hand, the timber trade can lead to greater net returns for forestry investments, and make sustained timber management as a prerequisite for stabilization of forest areas in the tropics more feasible in the long run. On the other hand, it appears that poorly managed and excessive timber extraction is leading to direct and indirect forest degradation in the short run. Chapter 6 will look in more detail at the domestic market and policy failures that are undermining the incentives for trade to contribute to sustainable forest management.

4

The Market for Tropical Timber Products

This chapter reviews current market conditions and trends in demand for tropical timber products.[1] Careful analysis of the factors underlying consumer demand is an essential pre-requisite to assessing the impacts of policy interventions in tropical timber markets, particularly policies which attempt to influence demand directly. Four key issues are addressed in this chapter:

- How have patterns of trade in tropical timber changed as a result of changes in the type and level of forest products consumption in major markets?
- What are recent trends in consumption, and how are they likely to affect tropical timber markets as a whole?
- What are the implications of the emergence of new markets, particularly those in developing countries that are not producers of tropical timber?
- How responsive are markets to changes in the prices of tropical timber products, what are the substitution possibilities for tropical timber products in international markets, and what are the implications for timber trade policy options?

INDUSTRIALIZED COUNTRIES

General market conditions in the major consumer markets and the role of tropical timber in total forest products consumption are briefly reviewed in this section. Particular attention is paid to the end uses of tropical timber, key determinants of demand, consumer preferences and technological developments.

Table 4.1 shows net imports of tropical timber by products and major importing regions. Japan is the dominant market for tropical timber, followed by Europe, then North America.[2] Over the last decade, Korea and Taiwan have emerged as

---

1 We are grateful for the assistance of David Brooks in our analysis of consumer markets for tropical timber products. His original report appeared as Annex C in Barbier et al (1993) and has been reproduced as Brooks (1993).

2 The data presented in Table 4.1 may reflect some 'double counting' – ie timber imported into Korea and Taiwan, processed and then re-exported to Europe, Japan or the US appears twice in the trade data. Therefore, the total volume of trade is likely to overstate the total volume of timber that actually enters the trade.

The Market for Tropical Timber Products   35

Table 4.1 Net imports of tropical timber products in major markets, 1990

Product	North America	Japan	Europe (EEC)	Republic of Korea	Taiwan[a]
	000 m^3 (product basis)				
Logs	3	11,319	3,303	3,731	4,193
Sawnwood	202	1,371	3,094	587	na
Veneer	24	117	229[b]	13	–
Plywood	1,099	2,718	1,030[b]	541	433
	000 m^3 (roundwood equivalents)[b]				
Logs	3	11,319	3,303	3,731	4,193
Sawnwood	364	2,468	5,569	1,057	na
Veneer	46	222	435	25	–
Plywood	2,528	6,251	2,369	1,244	996

Notes [a]Data for 1989
[b]Conversion factors (m^3 roundwood per m^3 product) are: sawnwood, 1.8; veneer, 1.9; and plywood, 2.3
Sources ITTO (1991); SBH (1991); WSC (1991); Foreign Agricultural Service (1990)

significant importers of tropical timber – by 1990 the level of imports by Korea and Taiwan were just below that of Europe. Imports by Korea and Taiwan are mainly destined for their export-oriented timber processing industries, with North America and Japan as the major end consumers of these 'secondary processed' products.

Japan continues to be the world's largest importer and consumer of tropical timber, accounting for roughly 50 per cent of tropical timber imports by the industrialized countries and more than 20 per cent of world trade in tropical timber products in 1990 (Table 4.1). Although Japan's domestic timber resources are substantial (25 mn ha of forests containing around 3 bn m^3 of timber, of which approximately 33 per cent is hardwood) imports of raw material and timber products into Japan account for nearly 75 per cent of domestic demand. Import-dependency in Japan has increased steadily since the mid-1970s, partly because the bulk of domestic timber resources are immature or uneconomical to harvest, and partly as a result of explicit trade and industrial policies. In recent years, 'secondary producers' of tropical timber products have increased shipments to the Japanese market.[3] For example, Japanese imports of furniture and furniture parts based on tropical timber have more than doubled in the past decade, with nearly all this growth accounted for by imports from secondary producers.

Table 4.2 summarizes the importance of tropical timber in the Japanese forest sector in 1990. In terms of both value and volume (all products expressed in a equivalent basis), log imports still account for the majority of Japan's tropical

---

3 'Secondary producers' are countries that import tropical timber in raw form, process it domestically and then re-export it. These are mainly the newly industrializing countries, such as Korea and Taiwan, which do not have substantial stocks of tropical timber but are competitive in certain types of wood processing.

Table 4.2 Importance of tropical timber in the Japanese forest sector, 1990

Type	Logs	Lumber	Plywood[a]
Tropical			
Hardwood	11,300	1,371	2,718
Softwood[b]	227	308	2
Temperate			282c
Hardwood	1,017	343	50
Softwood	16,455	7,061	
Total			
Hardwood	12,317	1,714	3,000
Softwood	16,682	7,369	52
Total	**28,999**	**9,083**	**3,095[d]**

Notes [a]Converted from square metres (m^2) at 135 m^2 per m^3
[b]Softwood volume from South Seas and 'other'
[c]Hardwood plywood not classified as tropical; Indonesia is the primary supplier
[d]Includes plywood classified as 'other'
Source WSC (1991)

timber imports. Tropical logs are used primarily in the domestic plywood industry, although their relative importance is lower than at any time in the recent past. Supply restrictions in tropical timber production, competition from product imports and a slow-down in construction activity (1991–2) have all contributed to the long-term decline in Japanese plywood production – which is now at its lowest level since the mid-1960s (Japan Lumber Reports 1992).

Housing and construction uses dominate *end use markets* for tropical timber in Japan, although tropical hardwood lumber (whether imported or manufactured in Japan) lost market share in all end use markets over 1980–9 (Ward Associates 1990). An important indicator of demand for wood producers in Japan is residential construction, which is closely correlated with economic activity and business cycles. Throughout 1970–85 total housing starts declined, as did the size of houses built and the share of new houses that are wooden. Total consumption of tropical lumber declined by more than 2 mn m^3 between 1980 and 1989. Although government intervention in housing markets and product promotion by US manufacturers (focused on construction methods) provided a boost to demand in the late 1980s, this tended to favour the consumption of softwood lumber. The recession in the Japanese economy in the early 1990s was again reflected in a decline in construction and housing starts, leading to a further decline in timber products consumption.

Changes in the mix of raw materials used are driven only in part by availability. For Japanese manufacturers, supplies might expand through use of non-traditional tropical species, or through use of non-tropical timber. In either case, we expect to see scarcity reflected in relative prices. Vincent et al (1990) found little statistical evidence for price-based substitution of lesser-known species (LKS) for well-known species (WKS) of tropical logs. However, there was some statistical support for the hypothesis of price-based substitution of temperate for tropical species in lumber manufacturing (Vincent et al 1991). Ward Associates (1990) report that tropical LKS are increasingly being used in both lumber and plywood manufacturing. Nevertheless, nearly all Japan's tropical log imports continue to come from Asia. Africa and Latin America supply very small quantities of logs to the Japanese market (Ward Associates 1990). Imports of tropical products are considerably more diversified, however.

Uncertainty concerning supply (or, at the very least, the process of changing sources of supply from domestic production to imports) is likely to have contributed to this decline in timber consumption, along with price-based substitution. For example, Ingram (in press) reports that after a period of declining real prices beginning in the late 1970s, real (deflated) prices of tropical sawnwood increased sharply in the late 1980s.

Perhaps more significant than an increase in the real price of tropical timber, however, is the change in prices of tropical timber relative to prices of substitutes (see 'Consumer prices, substitution and demand' below and Brooks 1993 for more detail on substitution). Figure 4.1 illustrates one measure of relative prices, based on data contained in a recent analysis of the Japanese market by Ward Associates (1990). The figure shows a rising trend in the ratio of the real price of tropical (luan) sawnwood to the real price of temperate (hinoki) sawnwood, over the period 1970–91. This trend may help to explain reductions in the market share of tropical timber reported by Ward Associates (1990).

Europe accounts for around 50 per cent of world forest products imports and 20 per cent of global tropical timber imports (Table 4.1, ITTO 1990). However, Europe is considerably more self-sufficient in forest products than Japan – European countries as a whole rely on imports for approximately 15 per cent of total consumption, of which tropical timber accounted for less than 20 per cent on a volume basis (ITTO 1990). Thus, tropical timber accounts for roughly 3 per cent of Europe's consumption of industrial timber products. However, tropical timber products are important to the European timber market and occupy a significant high-value niche in this market (ITTO 1990). Six countries account for 90 per cent

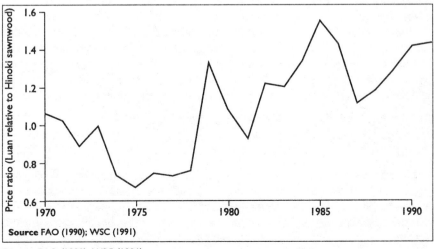

Source FAO (1990); WSC (1991)

Figure 4.1   Relative prices of sawnwood in Japan: Luan and Hinoki

Table 4.3 End uses tropical sawnwood and plywood in selected European countries

	Imports of tropical logs			Imports of tropical	
	Total	Used for lumber	Used for plywood	Lumber	Plywood
	000 m³				
Volume	11,300	2,614	8,684	1,371[c]	2,718[d]
	%				
Share of supply[a]	24	5	88	5	26
Share of imports[b]	39	11	92	15	88

Notes [a]Tropical share of total log supply (from domestic production and imports), and tropical share of total lumber and plywood supply
[b]Tropical share of total log, lumber, and plywood imports (including softwood logs)
[c]Includes hardwood lumber from South Seas and a small quantity of hardwood lumber from sources other than the South Seas, North America, Russia, Chile and New Zealand
[d]Tropical hardwood plywood imports (WSC 1991) converted from m2 (366.9 million m²) at 135 m² per m³
Sources Calculated and estimated from data reported by JAWIC (1991), WSC (1991), and Japan Lumber Reports (1992)

of total EEC consumption of tropical sawnwood and plywood: the United Kingdom, France, Germany, Italy, Netherlands and Belgium. Of these, the United Kingdom is the largest single consumer of tropical timber – 18 per cent of total European sawnwood and 37 per cent of plywood consumption in 1988 (Cooper 1990). As a result of proximity and historical ties, Africa supplies the majority of European imports of tropical logs. However, sources of tropical sawnwood and plywood are more diversified and include producers in Asia and Latin America.

Approximately 50 per cent of tropical log imports by the EEC are used to manufacture plywood, 33 per cent are used to manufacture sawnwood and the remainder are used to produce veneer (Cooper 1990; ITTO 1990 and 1991). However, patterns of end use of tropical sawnwood and plywood are not homogeneous across European countries. Table 4.3 shows that France, Germany and the UK differ in types of imports, sources of imports and patterns of use. For example, the UK relies almost entirely on imports of processed products while France and Germany have relied heavily on log imports in the past. French imports of tropical logs are supplied almost exclusively by African producers, while German imports are primarily from Asia.

Recent trends in European timber markets are broadly similar to those in Japan:

- a reduction in imports of tropical logs from the early 1970s
- a relative increase in imports of processed tropical timber products, and
- an overall decline in total tropical timber consumption.

These trends are expected to continue in the near future (ITTO 1990). For a number of countries and specific end uses markets may be static or even declining (Cooper 1990; ITTO 1990). Competing products, and to a lesser extent changing tastes and preferences, will lead to loss of market share or absolute reductions in quantities of tropical timber products consumed.

Canada and the US are relatively small participants in tropical timber trade, with a combined import value of tropical logs, sawnwood, veneer, and plywood totalling about US$ 550 mn in 1990. The US dominates the North American market, accounting for more than 90 per cent of total net imports into this region in 1990 (see Table 4.1). Tropical plywood is the main solid wood product imported into this region, accounting for nearly 90 per cent by volume (and more than 75 per cent by value) of imports. North America also imports substantial quantities of other tropical manufactured forest products, such as pulp and paper products. Indonesia, Malaysia and Brazil supply most North American imports of tropical timber. Imports from Africa are negligible.

As in the European markets, relatively high-value end uses of tropical lumber and plywood dominate the market. Nevertheless, tropical timber products face strong competition in nearly all end uses (Ward International 1992). Recent declines in construction activity (1990–1) reduced overall demand for all wood products, including tropical timber. However, temperate hardwoods, other wood-based products (such as particle-based panels), and other materials present significant competition for tropical timber (Ward International 1992). Trends in relative prices for imported hardwood plywood in the United States show patterns similar to those for tropical lumber in Japan. That is, increasing relative prices that can be expected to discourage consumption of tropical plywood, and encourage consumption of substitutes.

TROPICAL DEVELOPING COUNTRIES

Recent projections of consumption of forest products in developing and developed countries show that the anticipated growth in consumption in developing countries will exceed that in developed countries (FAO 1991). In many cases, projected growth rates for developing countries are *double* those of the developed countries.[4] In 1975 consumption of industrial roundwood and sawnwood in developed countries was nearly 5 times that in developing countries. By 1990 the ratio of consumption in the two groups was roughly 3 to 1, and in two decades the ratio is projected to be roughly 2 to 1.

Although the implications of the FAO projections may seem dramatic, they originate in two simple facts. First, when a percentage is calculated against a small base quantity, rates of increase are exaggerated. The level of total industrial timber consumption in tropical developing countries is small, both in absolute terms and relative to consumption in developed countries, hence rates of increase seem very high by comparison. Second, the FAO projections rely largely on forecast population growth (as well as income growth) to estimate future consumption of forest products. Higher population growth rates in developing countries lead to expectations of higher rates of increase in all aspects of consumption in developing countries. In contrast to the 'mature' markets of Europe and North America, which are characterized by relatively low rates of economic and population growth, the

---

4 Because developing countries are largely – although not exclusively – located in the tropics, it is reasonable to make inferences about timber product demand in tropical timber producing countries from developing country statistics.

developing countries are generally undergoing rapid change and development, with high rates of population and economic growth. As a result, over the next two decades developing countries are likely to be the fastest growing markets for tropical timber and timber products.

The implications of these trends for tropical timber producing countries must be examined closely. Growth in consumption of industrial products adds to pressure on tropical forests that already account for nearly half of world timber harvest (including fuelwood). Based on both the scale of consumption, and the mix of products, it is reasonable to ask if tropical countries are capable of meeting their projected domestic demand with available domestic resources. The world's industrial timber economy is predominantly based on temperate zone coniferous species. In spite of the fact that the tropical timber share of industrial roundwood harvest has increased significantly in the past four decades, in 1990 tropical timber accounted for less than 15 per cent of total *industrial* roundwood production (see Table 2.3a). To the extent that increased consumption of forest products in tropical countries follows the pattern of the industrialized forest economies of the temperate zone, domestic resources, although abundant in many cases, will not match the mix of raw materials required by technologies in production and consumption. As a result, increased imports of forest products will be necessary to satisfy consumption requirements and many developing countries will become net importers of timber products (see Chapter 2).

Figures 4.2 to 4.5 compare trends in production and consumption for four categories of industrial forest products, for tropical developing countries as a whole. Analysis of these trends provides some insights into the developing country timber products market:

■ The growth in both production and consumption of non-coniferous sawnwood has increased steadily from 1970 to 1990 at a consistent rate (Figure 4.2).
■ Net exports of non-coniferous sawnwood have changed little over the 20 year period, but wood-based panel exports have increased significantly (Figures

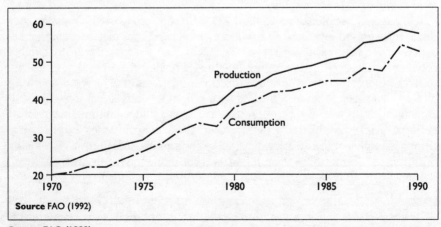

Source FAO (1992)

Figure 4.2 Production and consumption of non-coniferous sawnwood in tropical countries (mn m³)

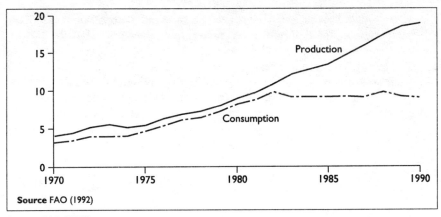

Figure 4.3 Production and consumption of wood-based panels in tropical countries (mn m³)

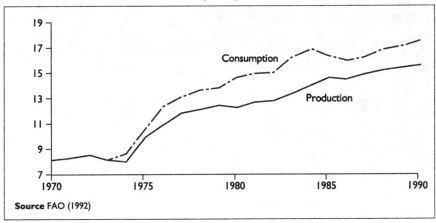

Figure 4.4 Production and consumption of non-coniferous sawnwood in tropical countries (mn m³)

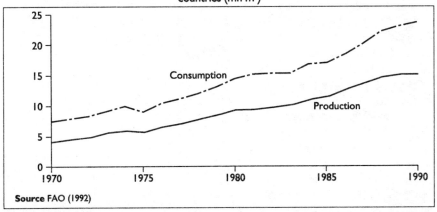

Figure 4.5 Production and consumption of paper and board products in tropical countries (mn mt)

4.2 and 4.3). This reflects the growth of a substantial, export-oriented industry. However, the failure of consumption to recover fully after the 1981–2 world recession is more difficult to explain.
- Consumption of coniferous sawnwood in tropical countries grew least quickly of all four product categories (Figure 4.4). Nevertheless, consumption growth was considerably higher than in developed countries and although growth in production was also significant, net imports are substantial.
- Consumption of paper and board products by tropical countries is increasing rapidly and is highly dependent on imports (40 per cent of consumption).

CONSUMER PRICES, SUBSTITUTION AND DEMAND

An understanding of how markets respond to changes in the prices of tropical timber and the availability of substitute products is essential in formulating appropriate timber management and trade policies. The conditions determining these responses and substitution possibilities include:

- the general market conditions determining the demand for tropical timber products in consumer markets;
- the effects of increases in the prices of tropical timber products on this demand;
- the extent to which other (eg temperate or softwood) timber products substitute for tropical timber products in consumer markets;
- the extent to which non-timber products compete with timber products in various end uses; and,
- the extent to which tropical timber products from one country or region substitute for products from another country or region.

The general market conditions affecting demand in the major consumer markets were covered earlier in this chapter. Here, we concentrate mainly on the latter points.

To some extent, the nature of market demand for tropical timber products can be described in terms of various standard measures of *price elasticity*. Price elasticities measure the relationship between a change in the quantity of a good that is consumed or imported and a change in its price, or in the price of a rival good. Thus, corresponding to each of the conditions noted above there is an elasticity measure.

The rest of this section discusses what is currently known about each demand and substitution effect by focusing on the results of various empirical estimates of the corresponding elasticity measures. Until recently, there were few studies of any of these effects, and most current studies generally focus only on the *own-price* elasticity of demand for tropical timber products.

Price Elasticity of Demand

As outlined by Constantino (1988) the (own) price elasticity of demand for tropical timber products is an important indicator in three general respects:

- A low (much less than one) elasticity measure indicates that, if the export price of the timber product increases, the quantity demand internationally would decline little; alternatively, if exports were increased, the international price would decline substantially.
- A high (much greater than one) elasticity would obviously have the opposite effects: an increase in export prices would have a large impact on the quantity demanded in international markets or, if exports were increased, prices would not fall substantially.
- For an individual producer country, a low elasticity of demand for its products implies a small international market relative to its level of exports, whereas a high elasticity indicates a large market relative to its exports.

Past empirical studies suggest that the own price elasticities of demand for tropical timber products are generally on the low side (see Table 4.4). In other words, consumers do not respond dramatically to price changes. Similar results are obtained in more recent analyses of specific regions and countries, notably SE Asia (see Table 4.5). For SE Asia in general, it appears that the elasticities of demand for log exports have been generally higher than for plywood and sawnwood exports. [5] It is not surprising therefore, that these countries have moved to restrict log exports in favour of expanding exports of processed products, notably plywood.

However, what holds true for specific regions or countries may not be applicable for the global market in tropical timber products. A recent analysis of the international demand and supply for non-coniferous logs, sawnwood and plywood by the world's major exporters and importers of these products has been conducted by Buongiorno and Manurung (1992). The resulting long-run own-price elasticity estimates and adjustment lags are shown in Table 4.6.

The elasticities of demand are generally small – only the demand for plywood has an elasticity greater than one. Although not all the products analysed by Buongiorno and Manurung were of tropical origin, the results contrast sharply with studies of the SE Asian region: demand elasticities for logs in the world market appear to be lower than for sawnwood, which in turn are less than those for plywood.

One explanation for the very low elasticity of global demand for non-coniferous logs is that over the 1968–88 period many major tropical timber producers implemented policies to restrict log exports (see Chapter 7). Although the fall in log exports by these producers was compensated by increased exports by other producers, the overall effect was that the overall supply of world exports grew slowly relative to demand. On the import side, this would mean that real prices of non-coniferous logs would rise faster than the quantity imported, which would translate into a low long-run elasticity of import demand. Such a scenario suggests that

---

5   In Table 4.5, the elasticity of demand by importers of tropical sawnwood of –5.67 calculated by Cardellichio et al (1988) and used by Vincent (1992) for Malaysia seems way out of line with other estimates. The earlier analysis by Vincent (1989) gives a more realistic estimate of –1.22 for Malaysian and SE Asian sawnwood import demand.

Table 4.4 Price elasticities of demand for timber

Source	Period	Countries	Products	Elasticity
Bourke	1988	Japanese imports from developing countries	Sawn timber	−1.3
			Veneer and plywood	−1.8
Brooks	1971–91	United States imports	Hardwood plywood	−1.2
Cardellichio et al	1965–87	N America/W Europe Japan Korea	Non-coniferous sawnwood	−0.5 −2.42 −1.06
		N America/W Europe Japan Korea	Coniferous sawnwood	−0.3 −0.67 −1.51
		United States Japan Korea	Non-coniferous plywood	−0.5 −0.55 −0.85
ECE/FAO	1964–81	Group I countries France Germany Netherlands Switzerland United Kingdom	Sawn hardwood	−0.46 −0.28 −0.34 – −1.07 0.49
		Group I countries	Sawn softwood	−0.58
		Group I countries	Plywood	−0.24
Kallio et al	1987	Countries with per capita income above US$ 3,000	Non-coniferous sawnwood	−1.2
			Coniferous sawnwood	−0.5
			Veneer and plywood	−0.4
Meyer	1952–75	Switzerland	Industrial wood	−1.4
NEI	1961–81	17 W European countries	Tropical hardwood	−0.34
Slangen	1963–81	Netherlands	Coniferous roundwood	−0.78
Wibe	1970–79	60 countries with per capita income above US$ 2,500 in 1975	Non-coniferous sawnwood	−1.19
			Coniferous sawnwood	−0.54
			Wood panels	−0.18

Sources Brooks (1993); Cardellichio et al (1989); and NEI (1989)

Table 4.5 Price elasticity estimates for SE Asian tropical timber

	Time period	Short run	Long run
Barbier et al (1994)			
Indonesia – log supply	1968–84		0.20
Indonesia – log export demand[a]	1968–84		–1.51
Indonesia – sawnwood supply	1968–88		0.27
Indonesia – sawnwood domestic demand	1968–88		–0.36
Indonesia – sawnwood export demand	1974–88		–0.68
Indonesia – plywood supply	1975–88		0.31
Indonesia – plywood domestic demand	1975–88		–0.91
Indonesia – plywood export demand[b]	1975–88		–0.46
Constantino (1988)[c]			
Indonesia – sawnwood export demand[d]	See note d	–0.08	–0.21
Indonesia – plywood export demand	1975–85	–0.26	–0.58
Importers – elasticity of substitution, tropical and temperate sawnwood	1975–85	1.30	2.11
Importers – elasticity of substitution, tropical and temperate plywood	1975–85	0.75	1.23
World – elasticity of substitution, source of origin, sawnwood	1979–85	2.50	4.39
World – elasticity of substitution, source of origin, plywood	1979–85	4.56	12.3
Vincent (1992)[e]			
Peninsular Malaysia – log supply	1973–89		1.1
Peninsular Malaysia – sawnwood supply	1973–89		1.7
Peninsular Malaysia – sawnwood domestic demand	1973–89		–0.55
Other tropical exporters – log export supply	1973–89		2.7
Other tropical exporters – sawnwood export supply	1973–89		0.7
Importers – tropical log demand	1973–89		–1.59
Importers – tropical sawnwood demand	1973–89		5.67
Vincent (1989)[f]			
Malaysia – log supply	1960–85		0.64
Malaysia – sawnwood supply	1960–85		1.00
Malaysia – plywood supply	1960–85		1.00
Malaysia – sawnwood domestic demand	1960–85		–0.27
Malaysia – plywood domestic demand	1960–85		–0.94
Other SE Asia exporters – log supply	1960–85		0.46
Other SE Asia exporters – sawnwood supply	1960–85		1.00
Other SE Asia exporters – plywood supply	1960–85		1.00
Other SE Asia exporters – sawnwood domestic demand	1960–85		–0.53
Other SE Asia exporters – plywood domestic demand	1960–85		–0.85
Importers – SE Asian and Malaysian sawnwood demand	1960–85		–1.22
Importers – SE Asian and Malaysian plywood demand	1960–85		–0.46

Notes: [a] Export price of Indonesian logs relative to average world price of temperate non-coniferous logs.
[b] Export price of Indonesian plywood relative to export wholesale spot price of Philippine (luan) plywood.
[c] Elasticity of substitution between sources of origin is defined as the percentage reduction in the ratio of imports from two different countries if the ratio of import prices of the two countries increases by 1%.
[d] Based on Buongiorno (1979) for coniferous sawnwood.
[e] Elasticities based on Cardellichio et al (1989). Note that in the latter study, Indonesian sawnwood and plywood supply are both estimated to have elasticities of 0.7, and Indonesxian sawnwood and plywood domestic demand have own-price elasticities of –0.92 and –1.55 respectively.
[f] Elasticities for sawnwood and plywood supply were assigned a value of 1 in the analysis

Source: Barbier et al (1994)

Table 4.6 Long term price elasticities of demand and supply across major tropical timber importers and exporters (1968–88)

	Elasticity	Adjustment lag (Years)
Import demand		
Non-coniferous logs	−0.16	6.7
Non-coniferous sawnwood	−0.74	4.6
Plywood	−1.14	4.1
Export supply		
Non-coniferous logs	0.70	5.8
Non-coniferous sawnwood	1.02	7.6
Plywood	0.20	6.0

Notes Although not all this trade was of tropical origin, 83% of the world imports of non-coniferous logs were from tropical sources, 62% for non-coniferous sawnwood and 75% for plywood.
Source Buongiorno and Manurung (1992)

over 1968–88 there were not many substitutes for non-coniferous (mainly tropical) logs.

On the other hand, the higher elasticities for non-coniferous sawnwood, and in particular, plywood imports would suggest that there may have been more substitutes available for these products in importing markets. This is not surprising, given that many of the major importing countries of these products also produce their own processed products, especially plywood, from either the non-coniferous logs they import or from timber (largely coniferous) which they produce themselves.

Table 4.6 also indicates that over 90 per cent of the adjustment of tropical timber product imports to changing prices occur within four to seven years. Relatively slow adjustment of log imports to price changes probably reflects the longer lead time required to adjust primary and secondary processing capacity in importing countries.

Substitution by Other Wood Products

As indicated above, an important factor determining the demand for tropical timber products is the extent to which other types of wood products, such as temperate (mostly softwood) products, are potential substitutes. This effect is commonly measured by the *elasticity of substitution* of products, which measures the degree to which one commodity (eg tropical wood products) can be substituted by another (eg temperate sawnwood).

For example, the elasticity of substitution for tropical and temperate products in consumer markets would indicate the extent to which the markets for these two types of products are interrelated, or whether there are essentially two different markets for two distinct commodities. This has important implications for exporters of tropical timber products. As noted in Chapter 2, the global markets for temperate (mainly softwood) products are currently much larger than for similar products produced by tropical countries. If there is a *high elasticity of substitution* between these two types of products, then an expansion of exports of tropical timber products could lead to expanded market share and increased export earnings – at the expense of the temperate product market. On the other hand, a high elasticity of substitution would also mean that any increase in the export price of tropical timber exports would lead importing countries to substitute with temperate wood products.

Table 4.5 presents the short and long run estimates of elasticity of substitution between tropical and temperate sawnwood and plywood calculated by Constantino (1988) for certain importing countries.[6] The results indicate that the elasticities of substitution between temperate and tropical wood products are significant but relatively low. This suggests that markets for tropical and temperate timber products are somewhat distinct; hence tropical producers of these products may have some difficulty in penetrating the larger temperate market. This conclusion is supported by market trends, which indicate that the market share of tropical plywood in the global Japanese and US plywood markets has remained essentially constant over time – in spite of the rapid rise in exports from tropical countries, especially Indonesia (Constantino 1988).

Table 4.7 A model of United States demand for hardwood plywood imports (1971–91)

Equation: $Q = \alpha P_m^{\beta 1} P_d^{\beta 2} E^{\beta 3}$

Variable	Description
Q	US imports of hardwood plywood, million square feet, surface measure.
P_m	Index of the unit value of hardwood plywood imports, deflated using the all-commodity producer price index.
P_d	Hardwood plywood and related products price index, deflated using the all-commodity producer price index.
E	Index of end-use activity, constructed from an index of the real value of private construction, the index of total manufacturing production, and the index of furniture production.

Coefficient estimates (t-statistics)

Constant (α)	Own-price elasticity ($\beta 1$)	Elasticity of substitution ($\beta 2$)	End-use coefficient ($\beta 3$)	R^2	D-W[a]
8.0256 (144.09)	−1.1573 (6.66)	1.4514 (5.45)	0.8293 (3.07)	0.83	1.37

Notes [a] Durbin-Watson statistic
Source Brooks (1993)

For this study a model of US demand for imported hardwood plywood was estimated for 1971–91, with the price of domestic hardwood plywood included as one of three explanatory variables (Brooks 1993). Table 4.7 presents the results, which match those of previous studies. Estimated own-price elasticity is slightly lower than other estimates, but still greater than 1. The price of domestic substitutes is

---

6 For sawnwood, the importing countries were China, Italy, Japan, the United Kingdom and the United States. For plywood, they were Japan, the Netherlands and the United States.

shown to be a relevant factor in US demand for imported hardwood plywood, although the effect is slight.

Other studies have examined the degree of substitution of temperate for tropical logs. For example, Vincent, Brooks and Gandapor (1991) sought to quantify the extent to which Japanese consumption of imported saw logs has shifted over the past two decades away from tropical and towards temperate logs. Of particular interest is how the increased scarcity of tropical hardwood saw logs induced by deforestation and export restrictions in SE Asia – Japan's main supplier – has affected Japanese demand for temperate softwood logs.

The analysis indicates that changes in Japan's imports of tropical hardwood logs were affected by both *price effects* and *technical changes* within Japan. The price effects suggest that the Japanese market for imported saw logs became more competitive, especially over the 1979–87 period, with the industry substituting more aggressively between tropical and temperate logs. Not surprisingly, however, substitution was found to occur more readily within a species group (eg among all temperate softwoods) than across groups (eg substitution of temperate for tropical timber). Imports of tropical saw logs were further reduced by technical changes in Japan, which outweighed any price effects. This may reflect efforts by the Japanese sawnwood industry to broaden its sources of logs, due to the imposition of export restrictions by tropical timber producers.

Diversification of sources of supply, including substituting softwoods for hardwoods, may be occurring in the South Korean market. A recent study of the market over 1970–90 found that the consumption of tropical hardwood logs, which are imported, is responsive to the price of domestically produced softwood logs (Youn and Yum 1992). However, consumption of softwood logs is not affected by the price of imported tropical logs. This suggests that in the Korean market there are strong possibilities for substitution of domestically produced softwood for tropical logs, but not the other way around.

Substitution by Non-Wood Products

Of course, tropical timber products may also be substituted by other, non-wood products in end uses. Although there is increasing anecdotal evidence of this occurring in many consumer markets, particularly in the construction and furniture industries, estimating the magnitude or scale of this effect has proved more difficult.

Alexander and Greber (1991) estimate the elasticities of substitution between softwood lumber and non-wood materials used in construction, using relative prices as an explanatory factor. Although the results are by no means definitive, there are indications of the type and magnitude of interactions among timber and other materials. A survey by the Netherlands Economic Institute (NEI 1989) also suggests that substitution between wood and non-wood materials may be significant. For example, plywood is believed to face competition from solid synthetic panels, with price strongly influencing the choice of product in the construction industry. However, statistical estimates of the elasticity of substitution between tropical timber products and non-wood materials are not available.

Substitution by Origin

The study of Japan by Vincent, Brooks and Gandapor (1991) discussed above indicated that there was substantial substitution between temperate logs from different regions. A further important issue is whether there is also significant substitution between tropical timber products from different importing regions or countries.

This substitution effect can be measured by the *elasticity of substitution between sources of origin*, which indicates the degree to which an importing country can substitute a tropical timber product from one exporting country with a similar product from another country. The measure is important because it provides an indication of potential *changes in market share*. For example, a high elasticity would imply that, at the expense of a competitor, an exporting country could increase its share of a market by reducing the relative price of its product. Alternatively, from the perspective of an importing country, any reduction in exports of tropical timber products from one country could be fairly easily replaced with imports from another – provided that the latter can increase its exports at prevailing prices.

Table 4.5 indicates the short and long run estimates of elasticity of substitution by origin for tropical sawnwood and plywood calculated by Constantino (1988) for certain importing countries. [7] The results show that the elasticities of substitution are very high, especially for plywood. This would suggest that the importing countries can substitute between sources of origin with relative ease. Moreover, the elasticities indicate how certain tropical product exporters – especially Indonesia – have been able to increase their market share of world tropical sawnwood and plywood exports very rapidly in recent years. For example, because of its large timber resource base, its abundant labour supply and its policy of restricting log exports, Indonesia has been able to expand its processing capacity, production and exports fairly quickly. [8] As a result, it has been able to undercut the price of its competitors and take over import markets where substitution between different product sources appears to be high.

Another indicator related to substitution between different sources of origin is the ease with which exporters can switch from one importing market to another when greater import restrictions are imposed in one of the markets. This effect indicates the degree of *trade diversion* internationally which may result from a market intervention that distorts prices and thus trade flows. The greater the ease to which exporters can shift from one import market to another in response to the trade intervention, the greater the degree of trade diversion.

---

7   For sawnwood, the data were imports from Brazil, Côte d'Ivoire, Indonesia, Malaysia, the Philippines and Singapore into Italy, the United Kingdom and the United States. For plywood, they were imports from Brazil, China, Indonesia, Malaysia, the Philippines and South Korea into the United States.

8   However, as will be discussed in Chapter 7, this has also resulted in processing inefficiencies, over-capacity, artificially low log prices and consequently greater pressure on timber resources.

Buongiorno and Manurung (1992) examine the effects of a 5 per cent import tax in the Union pour le Commerce des Bois Tropicaux (UCBT) of the EEC.[9] However, the trade diversion impacts of the tax varied significantly depending on whether it was imposed on tropical logs, sawnwood or plywood.

For example, although a 5 per cent import tax on logs would reduce imports in UCBT countries by about 1 per cent, imports in non-UCBT countries would not be affected. Because non-UCBT countries are by far the largest importers of non-coniferous logs, they are unaffected by a small tax imposed on the UCBT import market. As a result, the world trade in tropical logs changes very little, and exports and prices for producer countries remain the same.

In contrast, a 5 per cent import tax on tropical sawnwood into UCBT countries would reduce their imports by 2.2 per cent but also increase imports to non-UCBT countries by 0.3–0.9 per cent. For producer countries, this results in a net fall in quantities exported and prices by 0.7–0.8 per cent each. Net export earnings would therefore decline by US$ 16 mn.

Similarly, for plywood, a 5 per cent import tax would decrease UCBT imports by 3–5 per cent but increase non-UCBT imports by 1–1.5 per cent. For producer countries, exports fall by 0.2–0.3 per cent and prices by 1.2–1.3 per cent. The result is a decline in export earnings of US$ 25.5 mn, with Indonesia alone losing US$ 17.9 mn.

Illustration of the Total Demand Effect for Indonesia

Constantino (1988) illustrates the interaction of these different types of price and substitution effects in an analysis of the impact of a 1 per cent increase in the price of Indonesian plywood and sawnwood exports on consumer demand in Europe, Japan and the United States. Estimates are given for three sets of price elasticities:

- the elasticities of substitution between *sources of origin*;
- the elasticities of substitution between *tropical and temperate zone products*
- the *own-price elasticities of demand*.

Estimated values are provided for both the short run and the long run and for two categories of Indonesian hardwood exports: sawnwood and plywood (Table 4.5). Using this information and share-of-market data from published trade statistics, Constantino constructs a single price elasticity to measure the total effect of a change in the Indonesian export price on the international demand for Indonesian wood products.

As Constantino explains, an increase in the price of Indonesian timber exports will have three effects:

- Firstly, *foreign consumers will search for other sources of supply* of tropical wood products, eventually buying more from these other sources and less from Indonesia. However, even if alternative sources of tropical timber are

---

9 The UCBT countries are Belgium, Denmark, Italy, France, Germany, Luxembourg, the Netherlands, Portugal, Spain and the United Kingdom.

available, foreign consumers may not be able to substitute completely for Indonesian products. In this case, the Indonesian export price increase will lead to a rise in the composite price of all tropical products (proportionate to the importance of Indonesian products in total tropical exports).

- An increase in the composite price of all tropical timber products leads to the second predicted effect: ie *foreign consumers substitute products from temperate zones for tropical timber products*.
- The increase in the composite price of all tropical products will lead to an increase in the composite price of all timber products, tropical and temperate (in proportion to the importance of tropical products in total timber consumption in foreign markets).
- This in turn leads to the third effect of an increase in the export price of Indonesian wood products: *foreign consumers substitute other materials (eg plastics or metals) for timber*.

These three individual effects and their total effect are presented in Table 4.8, for the long run only. The elasticities used to calculate the tropical/temperate substitution effects and the source of origin substitution effects, as well as the share of Indonesian product in total tropical timber imports and the share of tropical timber imports in total timber consumption for each product category and each market, are taken from Constantino (1988). The (own-price) demand effect is calculated using price elasticities of demand estimated in the forest sector model of Indonesia developed for this book (see Barbier et al 1993a). These are somewhat higher than the estimates originally used by Constantino, although the resulting own-price demand effect is negligible in any case. Market share data is as above.

As indicated in Table 4.8, the most significant impact of an increase in the price of Indonesian wood product exports is that foreign consumers look for other sources of supply in SE Asia. This reflects a very high elasticity of substitution by source of origin. Some substitution of temperate wood products for tropical wood

Table 4.8 Long run impact of a 1% increase in the price of Indonesian hardwood exports in three foreign consumer markets (%)

Product	Importer	Own-price demand effect[a]	Tropical/ temperate substitution effect	Source of origin substitution effect	Total effect[b]
Plywood	Europe	−0.00	−0.03	−11.93	−12.0
	Japan	−0.06	−0.36	−7.17	−7.6
	USA	−0.03	−0.80	−3.44	−4.3
sawnwood	Europe	−0.0	−0.03	−4.30	−4.3
	Japan	−0.0	−0.03	−4.30	−4.3
	USA	−0.1	−0.17	−4.30	−4.6

Notes [a] Calculated from the following equation: $D = e_p \cdot s_p \cdot {}^1 s_{tp}$ where: e_p is the own-price elasticity of demand for plywood or sawnwood (from Table 4.5 above); s_p is the expenditure share of the tropical commodity in the total expenditure for plywood or sawnwood; and ${}^1 s_{tp}$ is the expenditure share of imports from Indonesia in the total expenditure for the tropical commodity.
[b] Calculated as the horizontal sum of the three component effects.
Source Constantino (1988) except own-price demand effect and total effect

products also occurs, but to a much lesser degree. Substitution away from timber products altogether is negligible.

If accurate, these figures would imply that Indonesia has little market power *by itself*, due to the ease with which foreign consumers can substitute other sources of tropical wood products. In other words, unilateral export price increases by any single tropical producer would simply reduce that country's market share. On the other hand, the relatively low estimated elasticities of tropical/temperate substitution, and of demand (own-price), indicate that tropical producers *as a group* may enjoy significant market power. This suggests that coordinated price increases by all tropical producers would not lead to a significant decline in total demand. Moreover, these figures also imply that a uniform levy on tropical timber imports by foreign consumers would not significantly reduce demand.

Constantino points out that his estimates of the elasticity of substitution by source of origin may be too high, since they are based on historical data during a period in which Indonesia was expanding plywood exports at the expense of other producers. Other tropical producers are assumed to be able to increase output in response to greater foreign consumer demand with no increase in unit costs. If competing tropical producers cannot easily expand production and export prices in other regions rise, then the source of origin effect will be reduced. Note that prices of wood products in the domestic Indonesian market are also assumed to remain constant. Hence the figures in Table 4.8 should be taken to represent an upper bound.

Willingness to Pay for Sustainably Produced Tropical Timber

The annex to this chapter provides additional qualitative evidence on the nature of consumer demand for tropical timber products and, in particular, the extent to which consumer awareness of the problem of deforestation may influence demand. Based on a review of previous surveys of consumers and manufacturers of tropical timber products in the UK market, and on preliminary results from a new survey of manufacturers carried out for this study, some indications of consumers' willingness to pay for *sustainably produced* tropical timber products are presented. The broad conclusions of this review are as follows:

- Surveys in the UK indicate that consumers *may* be willing to pay significantly more for tropical timber products obtained from sustainably managed forests. Indeed, some consumers have expressed an unwillingness to purchase *any* products containing tropical timber, due to its perceived impact on tropical forests.
- Manufacturers, on the other hand, do not appear to believe that their customers would be willing to pay as large a premium as the public claims, particularly in the current recession. Moreover, manufacturers believe that scope for a price premium exists only in the quality joinery and furniture markets. In markets where tropical timber products compete largely on price with temperate or non-timber substitutes, such as building and construction, manufacturers see much less room for a price premium on sustainably produced tropical timber.

The results of these surveys in the UK suggest some caution in assuming that consumers are willing to pay higher 'premium' prices for timber products from 'sustainably' produced sources. Such ambiguity may reflect confusion on the part of consumers from, on the one hand, well-advertised campaigns persuading people to boycott *all* tropical timber products, and on the other hand, competing and often unverifiable claims on the part of manufactures, retailers and even independent organizations to have 'certified' certain products as coming from 'sustainable' sources. As noted in the annex, the reluctance of UK consumers to pay a premium for 'sustainably' produced timber products may have been influenced by the prevalence of a major economic recession during the time of our survey. This undoubtedly had an effect on the rather pessimistic price outlook of manufacturers.

CONCLUSIONS

This chapter has reviewed the general demand conditions for tropical timber products, including a description of major markets and end uses, price trends and issues in substitution. The general conclusions of relevance to policy are:

- Japan will remain the predominant market for tropical timber, followed by Europe and North America. Types of product imports and end uses vary widely across different markets. Log imports still account for the majority of Japan's tropical timber imports, where the dominant end uses are in housing and construction. Product imports are more significant in both Europe and North America.
- Recent trends in major consumer markets show a reduction in imports of tropical logs only partly compensated by increased imports of processed tropical timber products. These trends are expected to continue, leading to loss of market share or absolute reductions in quantities of tropical timber products consumed.
- Growth of consumption of forest products in developing countries will exceed that in developed countries, due to higher rates of economic and population growth. Nevertheless, the developed countries will remain the larger markets in absolute terms, at least in the near future.
- Increased consumption of forest products in tropical producer countries will exceed available domestic resources, leading many developing producer countries to become net importers of timber products.
- Empirical studies suggest that demand for tropical timber products generally is not particularly price sensitive, ie own-price elasticities are low. Estimated elasticities vary among different product categories, however, with some authors reporting more elastic demand for log exports than for plywood and sawnwood exports, while others report the opposite.
- Studies of substitution between tropical and temperate timber products are scarce. Limited evidence suggests that the two groups are only slightly competitive, ie estimated cross-price elasticities are relatively low. The economic incentive to substitute away from tropical timber is increasing, however, as

tropical timber prices continue to rise relative to prices of temperate timber products.
- Limited anecdotal evidence suggests that substitution of non-wood products for tropical timber may be significant. For example, tropical plywood is reported to face stiff competition from solid synthetic panels, with price strongly influencing the choice of product in the European construction industry.
- Similarly, empirical studies show that the elasticity of substitution among different *sources* of tropical timber is very high, with plywood again most susceptible. This suggests that importing countries can substitute between sources of origin with relative ease but also that tropical producers can easily capture market share from their competitors by competing on price.
- These different estimates of price elasticities, if accurate, indicate that tropical producers *as a group* may enjoy significant market power. Coordinated price increases by all tropical producers would not lead to a significant decline in total demand. The limited evidence presented here also implies that a levy on tropical timber imports by foreign consumers would not reduce demand significantly.
- Some surveys indicate that consumers *may* be willing to pay significantly more for sustainably produced tropical timber products. Manufacturers are more cautious in their assessment of what level of premium the market would bear, particularly in more price sensitive markets such as building and construction materials.

Annex: UK Surveys of the Willingness to Pay for Sustainably Produced Tropical Timber

A number of organizations have carried out surveys of the public or of industry in the United Kingdom in attempt to determine whether and how much consumers might be willing to pay, over and above current prices, for tropical forest products originating from sustainably managed forests. A summary of the results of the different surveys is presented in Table 4.9.

Table 4.9 Willingness to pay for 'substantially produced' tropical timber

Study	Focus of survey/sample set/methodology	Results
Milland Fine Timber Ltd. 1990	Focus: propensity of manufacturers to pay more for sustainably produced tropical hardwoods Sample: 160 UK-based manufacturers Methodology: telephone interview	65 per cent willing to pay (WTP) more 26 per cent unwilling to pay more 9 per cent don't know of 65 per cent WTP more: 34 per cent WTP 5 per cent more 23 per cent WTP 6–10 per cent 5 per cent WTP 11–15 per cent 3 per cent WTP 16–20 per cent 4 per cent WTP 21–25 per cent 31 per cent don't know
World Wide Fund for Nature (UK) 1991	Focus: consumer willingness to pay for tropical wood products shown to have come from well-managed forests Sample: 2,100 members of the British public Methodology: personal interviews	13.6 per cent average willingness to pay above current prices
Friends of the Earth 1992	Focus: consumer concern about rainforest destruction Sample: 1,500 shoppers entering 'Do-It-Yourself' hardware stores in the UK Methodology: personal interviews	91 per cent expressed concern about rainforest destruction 58 per cent claimed that if they knew a timber product came from a rainforest they would not buy it
London Environmental Economics Centre 1992 (Annex D in Barbier et al 1993)	Focus: opinion of UK tropical timber importers on whether and how much their customers would be willing to pay for the assurance that any tropical timber they buy originates from sustainably managed forests Sample: 83 members of the UK Tropical Timber Federation's National Hardwood Importers Section Methodology: mail survey	71 per cent believe that their customers are not willing to pay more 25 per cent believe that customers are willing to pay between 5 and 10 per cent more

Milland Fine Timber (MFT) Limited, currently the sole UK distributor of the Ecological Trading Company's allegedly 'sustainable' timber, commissioned the Business Research Unit to conduct a survey in order to understand timber buyers' purchasing behaviour, the criteria used in selecting hardwood suppliers, the uses and knowledge of different hardwoods and buyers' propensity to pay more for sustainably produced hardwoods. Interviewed by telephone were 160 UK-based *manufacturers* operating in a variety of areas including joinery, cabinet making, and household furniture.

Of those interviewed 65 per cent were prepared to pay more for sustainably produced tropical hardwoods (73 per cent of household and office furniture manufacturers; 50 per cent of kitchen cabinet manufacturers); 26 per cent were not willing to pay more; and 9 per cent did not know. Only a third of those who claimed to be willing to pay a premium were prepared to pay more than 5 per cent above current prices. The main reasons expressed for not wanting to pay more were that tropical timber is already too expensive, that customers would be lost, that they could not afford it, and the concern that increased costs would be passed on to the customer. The sectors most likely to pay more according to this study are conservatory, household and office furniture manufacturers and cabinet makers.

The World Wide Fund for Nature (WWF) commissioned MORI to conduct a survey of public attitudes on tropical rainforests and the environment. A sample set of 2,100 representing all adults in Great Britain (excluding the Western Isles and Orkney and Shetland) were interviewed at their home. One of the main findings arising from this research was the claim that *consumers* are willing to pay on average 13.6 per cent more for wood products if they can be shown to have come from well-managed forests. Furthermore, one in six consumers interviewed claimed to have stopped buying tropical wood products because of their fears about continuing forest destruction.

Friends of the Earth (FOE), as part of a campaign targeted at major do-it-yourself (DIY) retail hardware chains in the UK, conducted a national survey of over 15,000 *members of the public*. People entering one of the major DIY chainstores (B&Q, Do-It-All, Great Mills, Sainsbury's Homebase, Texas Homecare and Wickes) were questioned on their concern over rainforest destruction. Of these, 91 per cent expressed a concern and 58 per cent said that if they knew a timber product came from the rainforest, they would not buy it. Of those surveyed 83 per cent agreed with the phrase that shops should 'face up to their environmental responsibilities and stop selling tropical timber products which contribute to rainforest destruction'. People were not asked how much they would be willing to pay for timber from sustainably managed sources. FOE concluded from these results that the public is concerned about the rainforest and the role played by the DIY stores and that, in their own rather provocative words, 'companies cannot continue to ignore what their customers want'.

For this book LEEC conducted a survey by mail of the 83 members of the U K Tropical Timber Federation's National Hardwood Importers Section.[10] The aim of the survey was to ask firms to estimate whether and how much their customers would be willing to pay over and above prevailing market prices for the assurance that any tropical timber they bought originated from sustainably managed forests. The LEEC survey thus differs from previous surveys in that it does not query consumers directly in order to determine willingness to pay (WTP) but rather elicits manufacturers' *perceptions* of consumer WTP. A 'single bid game' was used, ie survey respondents were asked to state the maximum price that they expected consumers would be willing to pay for timber from sustainably managed sources.

The survey also collected information from firms on their opinion of future trends in the trade of tropical timber and on how they are adapting to current political pressures on the trade arising from public concern about tropical deforestation. Additional information collected in this survey included:

---

10 We are grateful to Camille Bann for assistance in this survey. Further details on the methodology of this and the other surveys discussed in the annex can be found in Annex D of Barbier et al (1993b).

- the area of the trade the respondent was involved in
- the number of employees in the firm
- the type of timber imported and the percentage of each type by volume and by value
- major end uses of tropical timber
- the ease with which it was possible to obtain details from suppliers of the type of forest management employed in producing the timber they supplied
- whether the respondent would stock timber from both sustainable and unsustainable sources.

The number of responses received to the LEEC survey were small and thus must be interpreted carefully. Nevertheless, the replies received suggest that a large majority of respondants (17 out of 24 or 75 per cent) feel that their customers are *not* willing to pay a premium for tropical timber derived from sustainably managed forests. The main reasons expressed for this were the competitive nature of the industry and the present recession. A minority (6 out of 24 or 25 per cent) thought that consumers would be willing to pay between 5 and 10 per cent more. A single respondent felt that customers would pay up to 20 per cent more, but only for top quality tropical hardwood; any higher price would lead customers to use alternative woods. Of those who believed a premium could be levied, there was consensus that this was most likely to be achieved in the quality joinery and furniture markets and not at all within the construction industry.

Most respondents saw trade in tropical timber declining, at least in volume terms, over the coming years. A movement away from plywood towards further prepared and assembled products is anticipated. Firms see this as a way of promoting value added products in the countries of origin and thus aiding them to fund a long term forest policy and to earn income.

A number of respondents felt that the general public had been misinformed about the role of the timber trade in tropical deforestation, resulting in an adverse effect on sales. Generally respondents were sensitive to the problem of deforestation and the need to act responsibly to secure their position in the market place. Respondents claimed to be actively expressing their concern through support for the UK Forest Forever Campaign and ITTO guidelines for the year 2000, requesting documentation confirming sustainable supply from suppliers, issuing catalogues and mailshots on the issue and so on.

These preliminary results from the LEEC survey are at variance with previous surveys. They suggest that UK tropical timber importers believe, consumers are *not* willing to pay a premium for timber products originating from sustainably managed sources. This may reflect the current recession, the fact that the survey was addressed to manufacturers rather than to consumers or the nature of the sample.

In conclusion, some of the important points that emerge from these surveys of consumer willingness to pay for tropical timber products obtained from sustainably managed forests are:

- Surveys in the UK indicate that consumers *may* be willing to pay significantly more for tropical timber products obtained from sustainably managed forests. Indeed, some consumers have expressed an unwillingness to purchase *any* products containing tropical timber, due to its perceived impact on tropical forests.
- Manufacturers, on the other hand, do not appear to believe that their customers would be willing to pay as large a premium as the public claims, particularly in the current recession.

Moreover, manufacturers believe that scope for a price premium exists only in the quality joinery and furniture markets. In markets where tropical timber products compete largely on price with temperate or non-timber substitutes, such as building and construction, manufacturers see much less room for a price premium.

5
TROPICAL FOREST POLICIES AND THE ENVIRONMENT

Chapter 3 looked at both the direct and indirect impacts of tropical timber production and trade on forest degradation and its resulting environmental effects. Although these impacts may be significant, particularly in specific regions and countries, commercial logging is generally not the predominant cause of tropical deforestation. Other factors, such as agricultural land conversion, are much more important. Moreover, as much tropical timber production is increasingly destined for domestic consumption, it is difficult to apportion blame for *all* the environmental impacts associated with the logging of tropical forests on to the *international trade* in tropical timber products.

Nevertheless, to the extent that policies influencing the tropical timber trade undermine incentives for sustainable management of production forests, then the resulting forest degradation and environmental impacts will be 'excessive'. That is, the level of timber extraction and deforestation arising from commercial logging and trade is greater than what is deemed optimal from a social, and indeed perhaps even a global, point of view. As will be discussed further in Chapter 6, this situation essentially results from the failure of individuals in the market place and of governments to recognize fully and integrate *wider environmental* – or *non-timber* – values into tropical forest management, production and trade decisions. Unless these wider values are taken into account, harvesting levels may be *privately* efficient but fail to be *socially* optimal (efficient).[1]

Some of the environmental values lost through tropical timber exploitation and depletion, such as watershed protection, non-timber forest products, recreational values, may affect only populations in the countries producing the timber. Concerned domestic policy makers should therefore determine whether the benefits of incorporating these wider values into decisions affecting timber exploitation balance the costs of reduced tropical timber production and trade, as well as the costs of implementing such policies. The socially 'optimal' level of timber exploitation and trade is one where the additional domestic environmental costs of logging tropical forests are 'internalized' in production decisions, where feasible.

The world's tropical forests, including the remaining production forests, are also increasingly considered to provide important 'global' benefits, such as a

---

1   As discussed in Chapters 6 and 7, the many market and policy distortions in tropical forest countries often mean that harvesting levels are not even privately efficient.

major 'store' of carbon and as a depository of a large share of the world's biological diversity. Similarly, even some 'regional' environmental functions of tropical forests, such as protection of major watersheds, may have transboundary 'spillover' effects into more than one country. But precisely because such transboundary and global environmental benefits accrue to individuals outside the countries which exploit tropical forests for timber, it is unlikely that such countries will have an incentive to incur the additional costs of incorporating these 'global' values in their forest management decisions. Not surprisingly, sanctions and other interventions in the timber trade are one means by which other countries may seek to coerce tropical timber producing countries into reducing forest exploitation and the subsequent loss of environmental values. In addition, trade measures are increasingly being explored as part of multilateral negotiations and agreements to control excessive forest depletion, to encourage 'sustainable' timber management and to raise compensatory financing for tropical timber producing countries that lose substantial revenues and incur additional costs in changing their forest policy.

However well intentioned they may be, both domestic and international environmental regulations and policies that attempt to 'correct' tropical forest management decisions may have high economic, and even 'second order' environmental, costs associated with them. There is increasing concern that the potential trade impacts of environmental policies that affect tropical forestry and forest-based industries may increase inefficiencies and reduce international competitiveness. Moreover, the trade impacts of domestic environmental regulations may affect industries in other countries and lead to substantial distortions in the international tropical timber trade. The overall effect on the profitability and efficiency of forest industries may be to encourage tropical forest management practices that are far from 'sustainable'. Careful analysis of both domestic and international environmental policies affecting tropical forest sector production and trade is therefore necessary to determine what might be the full economic and environmental effects of such policies.

Finally, current import and export policies influencing tropical timber trade flows can themselves be a source of major environmental impacts in the sector. Protection of forestry and forest-based industries, whether explicit or implicit, can encourage inefficient expansion of domestic industries that leads to excessive tropical deforestation. Producer countries that are unable to penetrate protected international markets may be discouraged from developing efficient value-added processing, or alternatively turn to policies that subsidize processing in order to overcome import barriers. It is conceivable that the loss in value added leads to a higher rate of timber exploitation in order to increase earnings from raw log and semi-processed exports. On the other hand, over-expansion of processing capacity and inefficient operations could result in greater deforestation.

The long term price implications of existing and proposed global timber trade policies, including any barriers to trade in certain timber products, may affect the incentives for some producer countries to change their forest policy and management practices, and may even encourage an expansion of activities that are environmentally detrimental. In particular, as argued in Chapter 3, the failure to generate adequate returns from sustainable tropical timber management may mean that forest land is increasingly allocated to competing uses, such as agricultural expansion.

The above concerns over the links between timber exploitation and trade, forest depletion and environmental protection can be grouped into four key policy issues:

- the environmental effects of domestic market and policy failures through their impacts on tropical timber forest management;
- the environmental effects of current tropical timber trade policies pursued in both producer and consumer countries;
- the potential impacts of domestic environmental regulations pursued in both producer and consumer countries which affect the tropical timber trade; and,
- the scope for new multilateral or unilateral policy interventions in the tropical timber trade to assist in the promotion of more sustainable management of tropical production forests.

These four issues form the basis of the discussion for the following chapters.

6

DOMESTIC MARKET AND POLICY

As noted in the previous chapter domestic market and policy failures in tropical timber producing countries may drive a wedge between actual and socially efficient rates of timber extraction and trade. Removal or correction of these failures may provide incentives for sustainable forest management.[1]

THE EFFECT ON FOREST MANAGEMENT

Market failures can lead to poor forest management when markets do not operate efficiently or fail to reflect non-timber values. The presence of open access resource exploitation and public environmental goods, externalities, incomplete information, and imperfect competition can all contribute to the failure of the market in achieving a socially efficient level of forest management.[2] Usually some form of public or collective action, involving regulation, market-based (economic) incentives or institutional measures is required to put forest management on a sustainable footing – provided that the costs of correcting market imperfections do not outweigh the benefits produced.

If the policy interventions necessary to correct market failures are not taken, or over-correct or under-correct for the problem, *policy failures* can also exacerbate poor forest management. Policy failures also occur when government decisions or policies are themselves responsible for excessive environmental degradation. Linkages between policy and forest management may be direct or indirect. Forestry sector policies and their institutional and legal framework can have a direct impact on forest management. The elements of macro-economic policy – monetary, fiscal and international economic policies – have a more indirect impact on forest management.

Public policies influence forest management through their effect on the incentives for attaining:

- a *privately efficient* harvest level; and,
- a *socially efficient* harvest level by accounting for non-timber values.

---

1 For more on sustainable management techniques for forestry see ITTO (1990).

2 See Aylward, Bishop and Barbier (1993) and Hyde, Newman and Sedjo (1991) for further explanation and discussion of efficiency, rent capture and market failure in tropical forest management.

Most of the forest land in tropical timber exporting countries is effectively owned by the state (Aylward, Bishop and Barbier 1993). In most countries, however, the state entrusts certain management responsibilities for production forests to private firms through a system of concession agreements. Depending on the country, these responsibilities may include a combination of forest demarcation, protection, inventory, monitoring, silvicultural interventions. As private firms with management responsibility for forest areas, concessionaires are necessarily responsive to market signals and policy incentives that affect forest management. Depending on the nature of these signals and incentives, management will be driven either towards, or away from, privately and socially efficient harvest levels.

If a timber concessionaire is committed to a forest management plan that seeks to maximize the future value of timber production from the concession, the concessionaire can be said to be attaining a *privately efficient* harvest level. However, many concessionaires fail to design or carry out privately efficient timber extraction and forest management plans. Instead, they may opt to cash in on the timber value in the short run, and neglect to invest in management activities that would maximize the long run value attainable from the concession. As a result, the economic potential of the nation's timber resource is squandered due to economically inefficient forest extraction and management practices. For example, it has been estimated that the depreciation in value of Costa Rica's timber resources in 1988 and 1989 due to poor management cost the country 36 per cent more than the payments incurred on its public external debt (WRI and TSC 1991).

As noted previously, in addition to serving as a valuable source of timber, forest lands also provide a range of important non-timber goods and services. These may include: *non-timber forest products* such as rattan; medicinal plants and wild foods harvested by local people; and *ecological services* such as watershed protection, carbon storage and micro-climatic stabilization. Thus, the definition of economic efficiency can be expanded beyond that of a privately efficient harvest level – which includes only the efficient production of timber from the concessionaire's point of view – to a *socially efficient* harvest level that maximizes the full range of economic benefits that a tropical forest is capable of generating over time to society as a whole.

Barbier (1992 and 1993) suggests that an important factor in achieving socially efficient harvest levels is to take into account the tradeoffs between the production of 'competing' goods and services. If the forest land under concession yields important non-timber values, then a narrow preoccupation with attaining privately efficient timber harvest levels – whilst ignoring other, non-marketed non-timber goods and services – will fail to provide a nation with the most efficient use of its production forests. Ensuring that management of the production forest minimizes loss of these non-timber values, may be the most socially efficient approach.[3]

---

3   As discussed below, one approach to minimizing the loss of non-timber values is to assign 'protected area' status to forest lands that are important for their provision of watershed protection, biodiversity or other non-timber values. However, it is of course expensive for a developing country to protect its forest lands indefinitely. On the other hand, as Thailand and the Philippines have discovered in recent years, the overlogging of production forests can lead to serious watershed degradation problems.

An illustration of the concept of *privately and socially efficient* harvest levels is provided by Paris and Ruzicka (1991). Table 6.1 summarizes their 'guesstimate' of the private and social returns to selective logging on steeply sloped (30 per cent–50 per cent) old-growth forest in the Philippines. Selective logging yields positive private returns when the revenues from selling logs (ie the value of the log harvest) are set against the costs of roadbuilding, harvesting and transportation. After accounting for the costs of protection, timber stand improvement and enrichment planting to ensure the sustainability of production, the private returns are reduced, but still positive. However, when the loss of non-timber values is

Table 6.1 Private and social efficiency: logging in the Philuppines (Philippino Pesos (P) per ha per year)

	Sustainable Management Scenario[a]
Value of log harvest[b]	5,720
Roadbuilding, harvesting and transportation costs	−2,369
Net financial return to short run timber harvesting	**3,351**
Internalizing cost of sustainable timber management: Cost of protection, timber stand improvement and enrichment planting	−1,000[c]
Net returns to 'privately efficient' timber harvesting	**2,351**
Shadow pricing adjustments:	
Adjustments to market price of logs	572
Adjustment to local harvesting costs[e]	474
Social costs of degradation of non-timber values: Cost of marginal offsite damage to downstream activities	−6,245[f]
Net returns to 'socially efficient' timber harvesting[g]	**−2,848**

Notes [a]Old growth forest selectively logged and subsequently protected.
[b]Legal operations using existing selective logging systems assumed. Private profits of illegal operators will be higher. Different combinations of yield and price are possible to capture the variations in the quality of standing forest. Assumption is that 1 ha of old-growth forest of 30–50% slope sustainably yields 100m³ every 35 years or 2.86 m³ per year. Market price is P 2,000 per m³.
[c]P 1,000/ha for 1 year to ensure sustainability of production on the 1 ha in question.
[d]Equal to market price adjusted upwards by 10% to account for low cost illegal supplies.
[e]Standard conversion factor 0.8 applied.
[f]P 2,600 for 3 years discounted at 12%. Off-site damage assumed to be limited in duration instead of being sustained in perpetuity. It would appear that Paris and Ruzicka see this as the damage resulting from the selective logging of a single plot. It would also seem that the magnitude of the figure is associated with the rather steep slope on which the logging is conducted. This example, therefore, does not indicate that *all* logging of old-growth forest in the Philippines is unwarranted from the standpoint of social efficency.
[g]Paris and Ruzicka acknowledge that their efforts are intended to be only a rough indication and are based on estimates prepared in 1990 as part of the formulation of the Philippine Master Plan for Forestry Development.
Source Paris and Ruzicka (1991)

taken into account the net economic gain is negative. In this case the magnitude of the estimated damage to downstream activities indicates that the Philippines would be better off not harvesting old-growth forests on such steep slopes. However, the existence of market and policy failures drives a wedge between the privately and socially efficient harvest levels. Therefore, whilst it is in the direct economic interest of the private concessionaire to continue unsustainable harvesting on the steep terrain, it is not in the interests of the Philippines.

Forest land use choices can also be affected by policies in other *sectors* of the economy and by general *macro-economic* policies. (The effects of macro-economic policies are discussed in more detail later in the chapter.) For example, an alternative use of forest land is agricultural cultivation. Agricultural pricing and investment policies can, therefore, influence the incentives for expansion of 'frontier' agriculture on to timber production forest land. As a result, managers of privately owned forest land feel the impacts of market signals and policy incentives in the agricultural sector as well as the forestry sector. However, concessionaires, who usually do not have the option of conversion (ie they have a concession on timber rights only), are less likely to feel directly the impacts of policies in other sectors.

In addition, the *institutional setting* within which a nation's land use policy is determined sets important limits on the use of market signals and policy incentives to encourage sustainable forest management. The designation of a nation's land as conversion, production or protection forest is likely to be a result of the nation's overall land use policy. If national land use policy dictates the conversion of certain forest areas it is unlikely that market signals or policy incentives in the forest sector will have much of an impact on the management of these forests. Thus, within a nation's land use framework there may be limits to the manipulation of incentives in order to secure sustainable management of all forests.

It is theoretically possible to determine an economically efficient land use policy that achieves a socially efficient allocation of land between these different competing uses. However, land use policy is usually not based solely on economic criteria, but is designed to meet a mix of economic, political and social objectives. This study takes the viewpoint that political and social objectives that enter into land use policy are often based on legitimate concerns – such as an equitable distribution of resources. These objectives will differ for each country and will necessarily be determined largely through the play of internal and external political and social forces governing the political process of each country. Nonetheless, when the land use planning process is driven by corruption, political favouritism or simply mismanagement – with no regard for either economic or equity concerns – the process itself may warrant the label of policy failure.

This discussion has provided a background to understanding how public policies may affect the incentives for timber extraction and forest management. Designing policies to improve forest management is clearly complex and requires careful attention to their potential direct and indirect effects on harvesting incentives. As the examples provided below illustrate, many domestic policies do not even begin to approximate the appropriate incentives required to achieve a *privately* efficient harvest level much less a *socially* efficient one. More often than not, pricing, investment and institutional policies for forestry actually work to *create* the conditions for short-term harvesting by private concessionaires, and in some instances, even *subsidize* harvesting at privately inefficient levels.[4]

In the following sections, the empirical evidence demonstrating the linkages between market failure, domestic policies and the incentives for sustainable forest management is reviewed.

Within the forest sector, Hyde, Newman and Sedjo (1991) suggest that *royalty, contract and concessional* arrangements are key incentive mechanisms determining levels of trespass, high-grading and loss of non-timber values.[5] Both trespass and high-grading usually indicate the failure of policy to internalize the private costs of long run forest management, thereby leading to a short run, privately inefficient harvest level. The loss of non-timber values is generally a result of market failure and leads to socially inefficient harvest levels. In addition, the failure of governments to collect excess profits from concessionaires may provide an additional incentive for timber operators to mine one concession after another in quick succession.

CONCESSIONS AND PROPERTY RIGHTS

Concessions typically involve a commitment on the part of the concessionaire to manage the concession over a given period of time. If the *length of the concession* does not cover at least the length of period required for one full rotation there will be little incentive for the logging company to manage the forest in a privately efficient fashion by developing a management plan aimed at maximizing the long run productivity of the concession, undertaking silvicultural measures that require investment in the short term, or minimizing logging damage. Short concession periods give the concessionaire little reason to be concerned about future revenue streams from the resource, leading to a rush to 'mine' the forest within the given concession period.

For example, throughout SE Asia the allocation of timber concession rights and leasing agreements on a short timescale, coupled with the lack of incentives for reforestation, have contributed to excessive and rapid depletion of timber forests (Barbier 1993). For many years concessions in the Philippines were awarded for 25 years but some for as little as 5. In Peninsular Malaysia 65 out of the 95 concessions in permanent and conversion forests were set for a period of 15 years or less. In general, Poore et al (1989, Chapter 5) found that concessions in Asia

---

4   For further examples and discussion see Aylward, Bishop and Barbier (1993); Barbier (1992 and 1993), Barbier, Burgess and Markandya (1991); Gillis (1990); Hyde, Newman and Sedjo (1991); Pearce (1990); Repetto (1990) and Repetto and Gillis (1988).

5   *Trespass* is a forestry term that refers to losses due to logging theft, which could also be extended to include losses due to graft. *High-grading* refers to the removal of high-valued timber leaving a degraded timber stand (Hyde, Newman and Sedjo 1991).

were issued for periods of 21 to 25 years, even though the minimum realistic felling cycle is 30–35 years and the rotation 60–70 years.

Poore et al (1989, Chapter 3) also suggest that concession size and duration are an important problem in attaining sustainable forest management in Africa. A review of policy in six African countries revealed that cutting cycles of from 15 to 25 years were the norm despite an assessment that in biological terms a 40 year cycle would be more appropriate. Although Ghanaian foresters suggest 5,200 ha as the minimum size for effective concession management, the government allocates concessions as small as 800 ha. In addition, these concessions are rarely offered for more than one cutting cycle, thus providing little incentive to the concessionaire to be concerned about future productivity. Grut, Gray and Egli (1991) report that the lengths of concessions for logging in the Congo are set at a maximum of 7 years. Côte d'Ivoire sets 5 year renewable concessions with 10 to 15 years allocated when exploitation is aimed at primary or secondary processing.

In addition, governments sometimes fail to control effectively encroachment on state forest lands. Left in a state of open access, the forest is quickly degraded as squatters seek the short term gains from mining the soil for agricultural cultivation. Cruz and Repetto (1992) indicate that excessive conversion of upland forest areas to agricultural use is occurring in the Philippines due to the government's unwillingness, or inability, to enforce its own regulations against conversion or to support traditional communal tenure systems. Three possible solutions exist in such cases:

- privatize the resource;
- devolve management responsibilities to local communities; or
- increase government capability to manage the resource.

Bhat and Huffaker (1991) examine the possibilities of privatizing government-owned forests in Karnataka, Western Ghats, India to combat open access over-exploitation caused by over-harvesting of forest foliage for use as a mulch for areca nut orchards. They found that under certain conditions privatizing the forests would fail to resolve the problem of forest degradation. Privatization may cause the new owners to pursue a strategy of 'mining' and abandoning the forest instead of following a recovery strategy leading to ecologically balanced use. Arnold (1990) argues against the privatization of state forest lands in India based on the four essential roles these common areas play in promoting rural livelihoods:

- Products flowing from the common areas complement resource flows from private property resources and are thus critical to their sustainability.
- Common areas are often a mainstay of the livelihood of the poor who cannot afford to develop private property.
- Common areas also tend to benefit from economies of scale in management that would not be achieved by individual management.
- Active collective management of common areas assists in maintaining local institutions and management skills.

Thus, although open access conditions may cause extractive activities that threaten forest resources, a simple solution of internalizing the costs by making the forest user the forest owner may not guarantee the continued existence of the forest base. Other policy responses may be more appropriate, for example encouraging communal management of the open access forest resources. By assigning property or usufructuary rights to a communal group, incentives may be provided to counteract the degradation brought on by the open access management of forest resources.

Devolution of management responsibility is, of course, not the only means to provide incentives for privately and socially efficient management of forest resources. One alternative is simply to improve the existing capability of forest departments to oversee the country's forest resources. A key element in such a strategy would be to upgrade personnel, facilities and equipment – all of which costs money. Chapter 10 discusses the funds that may be required and how such funds could be generated from the timber trade and consumer countries. Another possibility – to which we now turn – is how best to increase the government's capture of the excess profits garnered by concessionaires.

FOREST PRICING AND ITS ADMINISTRATION

The economic rent (ie the net returns) of harvesting activities is represented by the excess of the market price for timber over the costs of attaining a socially efficient harvest level – assuming that the loss of any non-timber values are internalized. In practice, a socially efficient harvest level is rarely if ever achieved. Instead, as the 'mining' of the forest base proceeds a country is left with the decision of how to allocate the short-run financial profits of timber harvests. Meanwhile – as demonstrated earlier by the example from Paris and Ruzicka (1991) – the failure to take into account the additional social costs of the loss of non-timber values can lead to an over-estimation of the net returns to timber harvesting, and lead to socially inefficient harvest levels.

The question of to whom this rent is allocated is not strictly an economic issue. Hyde, Newman and Sedjo (1991) strongly suggest that the allocation of rent is merely a matter of distributive preference. Each country must make its own judgement over the equitable distribution of the nation's natural wealth. Such concerns over equitable distribution are, however, likely to be influenced by expectations concerning the ensuing productive use of such economic surplus. More importantly, if a government leaves all the profits to the concessionaires without seeking to appropriate a reasonable return as the owner of the resource, the government leaves a powerful incentive in place for concessionaires to 'mine' the forests. This is particularly true in cases where – as discussed above – concessions policy and regulation are not strictly controlled and enforced.

Vincent and Binkley (1991) note that stumpage prices (ie the prices of harvested logs at the stand) have a crucial role to play in the interrelated dynamics of timber reserve depletion and processing expansion, particularly in facilitating the transition of the forest sector from dependence on old-growth to secondary-growth forests and in coordinating processing capacity with timber stocks. Unfortunately, in most developing countries, stumpage prices tend to be administratively determined rather than set by the forces of supply and demand. These administrative prices tend to understate stumpage values, and thus fail to

Table 6.2 Saw timber supply elasticities for regions with endogenous models

Region	Coniferous		Non-coniferous	
	Stumpage	Delivered price	Stumpage	Delivered price
USA				
Western Washington and Oregon (Westside)				
Private landowners	0.7	1.5		
Public landowners	1.3	2.8		
Eastern Washington and Oregon (Eastside)				
Private landowners	1.1	2.3		
Public landowners	0.9	1.8		
Inland (Rockies and Inland California)				
Private landowners	0.8	3.0		
Public landowners	0.5	1.7		
South	0.7	1.0	0.5	1.2
North	0.5	1.4	0.7	1.1
Canada				
BC Coast		3.2		
Interior Canada		1.1		
Eastern Canada		1.5		
South America				
Chile		2.8		
Europe				
Finland	2.9	3.9	0.9	1.1
Sweden	0.4	0.5		
Rest of Western Europe	1.0	1.2	1.0	1.1
Asia				
Japan		0.9		
East Malaysia, including Sabah, Sarawak and Brunei			2.2	2.7
West Malaysia, including peninsular Malaysia and Singapore			0.9	1.1
Indonesia			0.9	1.0
Philippines			1.2	1.2
Papua New Guinea			1.6	1.7
Oceania				
New Zealand	1.0	2.2		

Source Cardellichio et al (1989)

reflect increasing scarcity as old-growth forests are depleted. A number of economic and environmental distortions may result:

- old-growth forests are depleted too rapidly;
- forest land is inappropriately cleared for agriculture or other uses;
- inadequate and inappropriate investment is made in second-growth forests;
- inefficient processing facilities are installed;
- decisions on log and lumber trade policies are inefficient and encourage unsustainable management practices; and,
- elaborate and counter-productive capital-export controls are needed to ensure that resource rents are not repatriated.

Low stumpage prices relative to harvest and delivery costs mean that the latter have a larger role in timber supply decisions. However, this does not mean that producers are insensitive to prices. Limited evidence suggests that the timber industry in tropical producer countries responds to prices in a similar manner to the timber industry in temperate producer countries (Cardellichio et al 1989). This is illustrated by the comparable range of values of the price elasticity of supply in both regions (Table 6.2).

Vincent and Binkley (1991) cite two studies – logging of old-growth forests in peninsular Malaysia and Ghana – that provide examples of how long term forestry policy and development can be derailed by poor forest pricing policy. In Malaysia, wood prices – timber charges, log prices and sawnwood prices – have been kept artificially low at every stage of forestry development, from log exporter to exporter of primary products to exporter of downstream products. The result has been the development of processing capacity that exceeds the forests' sustained yield capacity.[6] Like Malaysia, Ghana has shown little success in establishing plantations as an alternative source of timber to the natural forest. Artificially low royalty rates for natural forest timber mean that concessionaires' incentives to invest in plantations is limited and at the same time, through its impacts on delivered log prices, has helped encourage the over-expansion of domestic processing capacity.

Table 6.3 indicates government rent capture from tropical timber in five countries. By not charging sufficient stumpage fees and taxes, selling harvesting rights too cheaply or by failing to collect forest fees most tropical timber producing countries have allowed the resource rents to flow as excess profits to timber concessionaires and speculators, often through short-term harvesting operations. For example, if the Philippine government had been able to collect the full value of actual rents in 1987, its timber revenues would have exceeded US$ 250 mn – nearly 6 times the US$ 39 mn actually collected.[7] Instead, excess profits of at least US$ 4,500 per ha went to timber concessionaires, mill owners and timber traders (Repetto 1990).[8]

---

6 For more discussion on this point see Jonish (1992).

7 This, of course, assumes that the mechanism for collecting these profits would not induce concessionaires to reduce harvest levels, thereby lowering the amount of rents available.

Table 6.3 Tropical timber rent capture

	Potential rent from log harvest (1) (US$ mn)	Actual rent from log harvest (2) (US$ mn)	Official gov rent capture (3) (US$ mn)	3/2 (%)	3/1 (%)
Indonesia (1979–82)	4,954	4,409	1,644	37.3	33.2
Sabah, Malaysia (1979–82)	2,198	2,094	1,703	81.3	77.5
Philippines (1979–82)	1,505	1,033	171	16.5	11.4
Philippines (1987)	256	68	39	57.1	15.3
Ivory Coast	204	188	59	31.5	28.9
Ghana	–	80	30	38.0	–

Source: Repetto (1990); and Repetto and Gillis (1988)

According to Grut (1990), stumpage and other fees paid by concessionaires have not been increased for many years in most African countries. As a result, these fees have declined in real terms and are below what most private companies are willing to pay for access to timber reserves. Grut, Gray and Egli (1991) report that in Ghana, total forest fees represent only about one-sixth the potential forest revenues that could be charged by the government. Annual collection of forest fees by the government netted only US$ 0.5 mn (or (US$ $0.38/m^3$) while the projected capturable rent was US$ 2.98 mn. In the Ivory Coast, a World Bank appraisal found that stumpage fees at US$ $0.58/m^3$ bore no relationship to the actual value of standing timber. Forest fee collections in Cameroon in 1987 came to US$ $5.40/m^3$ representing just 2–4 per cent of the export value of the logs. A 1989 World Bank appraisal mission to Guinea, found that the stumpage fees varied between US$ 0.50 and $0.68/m^3$ representing just 1 per cent of the local market value of the wood. Stumpage fees had not been adjusted for 6 years.

Hyde, Newman and Sedjo (1991) point out the difficulty experienced by forestry departments, particularly in developing countries, in administrating and collecting various timber fees and taxes. Grut, Gray and Egli (1991) found that low collection rates, arbitrary allocation of concessions, lack of forest management and poor enforcement of legal concession practices are a number of administrative

---

8 As noted above, rent capture per se may not be as fundamental to an efficient outcome as ensuring proper 'internalizing' of the user costs of timber exploitation through appropriate contractual and concession terms. That is, even if government rent capture is low, it can still ensure through proper concession arrangements that the stand is harvested at the private long-run efficient level. However, in many countries poor rent capture and poor concession policies go hand in hand, combining to produce short-term and rent-seeking behaviour in concessionaires. See Aylward, Bishop and Barbier (1993) and Hyde, Newman and Sedjo (1991) for a more detailed review of efficiency and rent capture issues.

problems faced by West and Central African countries. In the Congo only about one-fifth of volume-based forest revenue is collected. Low collection rates are a result of problems with log scaling or measurement, unreported logs and illegal logging, under-measurement of processed products, and transfer pricing.

The Report of the Commission of Inquiry into Aspects of the Timber Industry in Papua New Guinea (PNG) calculated that in 1986 and 1987 transfer pricing on log sales meant that PNG lost US$ 27.5 mn in foreign currency earnings and that the government lost US$ 4.3 mn in company tax.[9] In addition, not a single logging company declared a profit in PNG until 1986. The Commission of Inquiry also found that misdeclaration of species is another form of disguising profits – in 1986, misdeclaration of species allowed one company to reap an extra US$ 300,000 in revenues (Asia-Pacific Action Group 1990).

Grut (1990) cites the possibility of generating additional economic benefits from wood production in Ghana by using pricing policy to alter the incentives for wood usage. A 1988 World Bank appraisal of Ghana's Forest Resources Management Project estimated that 33–50 per cent of the wood cut from tropical forests in Ghana is left to rot on the forest floor. Further processing in Africa by saw and veneer mills typically results in the loss of 50 to 60 per cent of the wood, whereas it should be possible to confine processing wastage to as low as 40 per cent. The appraisal mission reported that reductions in wood wastage brought on by a project to implement higher concession fees, and to provide technical assistance on logging and inventory would produce an economic rate of return of 25 per cent.

Eventually, Grut expects that improvements in the forest department's inventory function will allow the department to assess concession fees based on standing trees instead of the current method of levying fees on extracted timber at roadside. Such a pricing system would further reduce any incentive the concessionaire might have to discard valuable wood during the logging process. In addition, more precise pricing will gradually lead towards increased efficiency and improved recovery of processed woods such as sawnwood and plywood.

The excess profits available to concessionaires, combined with failures in concessions policy and its enforcement, often contribute to excessive rent-seeking behaviour in tropical timber production (Gillis 1990). Concessionaires are tempted to push quickly through their existing concession in order to open up additional stands, or even new concessions, for harvesting ahead of their competitors. In such a climate the incentives for graft and corruption are rife. Effort devoted to obtaining a licence to valuable concessions can become an enterprise in itself. The Barnett Report lists a series of bribes and other illegal and corrupt practices that have taken place over the years in the PNG forest industry (Asia-Pacific Action Group 1990). In the Philippines, almost all the large logging companies have senior politicians on their boards, and it is generally the politicians,

---

9   Transfer pricing occurs when a vertically integrated company records a below market price for timber it is exporting to its overseas operations. The profit that would have accrued to the company in the exporting country is thereby transferred overseas to the importing country at the point when the company either sells, or 'charges' itself, the going market rate for the timber.

not forestry officials, that ultimately determine concession policy and allocation (Poore et al 1989, Chapter 5).

Grut, Gray and Egli (1991) suggest that higher rent capture by the authorities would reduce 'rent seeking' activities necessary to acquire concessions and curtail the inefficient logging and processing that is currently engendered by low input prices paid by timber companies. Grut (1990) suggests that projects within forest departments aimed at raising such fees will yield a high financial rate of return and enable forest departments to expand and improve underfunded forest management tasks.

Much of the problem may have to do with the complexity of fees and concession arrangements, which makes enforcement and supervision of revenue collection difficult. In their review of forest pricing and concession policies in West and Central Africa, Grut, Gray and Egli (1991) suggest replacing the multiplicity of forest fees with an annual concession rent, set by competitive bidding, and replacing logging concessions with forest management concessions that should be regularly inspected. In suggesting a similar scheme – a competitively bid lump sum fee for harvest rights to a particular area – Hyde, Newman and Sedjo (1991) indicate that one advantage of such bidding systems is all the risks associated with pricing are shifted on to concessionaires. This solves the problem of loading forest departments with the tasks of determining fees and pricing. As a result, the role of the forest department in managing production forests assumes more of an emphasis on monitoring of concession management.

MACRO-ECONOMIC POLICY

The impact of policy upon forest management is not limited to forest-specific sectoral policies. As indicated earlier in this chapter and in Chapter 3, general land use policies, as well as agriculture and other sector policies will have an impact on forest management by altering the relative incentives for converting land to alternative uses. In addition, general macro-economic policies may also influence forest management. Such macro-economic policies – excepting trade policy which is left for Chapters 7 and 8 – are discussed below. Figure 6.1 summarizes the macro-economic and sectoral instruments that can be used to provide incentives for sustainable forest management in tropical timber producing countries.

Macro-economic policy – be it international economic, fiscal or monetary policy – may have an impact on forest management through its effect on the timber trade. For example, an overvalued exchange rate acts as a subsidy to urban consumers on imported goods, whilst implicitly taxing timber exports produced domestically. Real currency devaluations, as frequently required by structural adjustment programmes for indebted developing countries, remove existing distortions and provide incentives for greater domestic production of exportables, including timber products. This is due to increasing international price competitiveness and increased domestic demand for home produced goods as imported substitutes become relatively more expensive. Both impacts can directly encourage socially inefficient harvest levels through the expansion of wood production for the international and domestic markets. Increasing pressure to meet debt serv-

Domestic Market and Policy 73

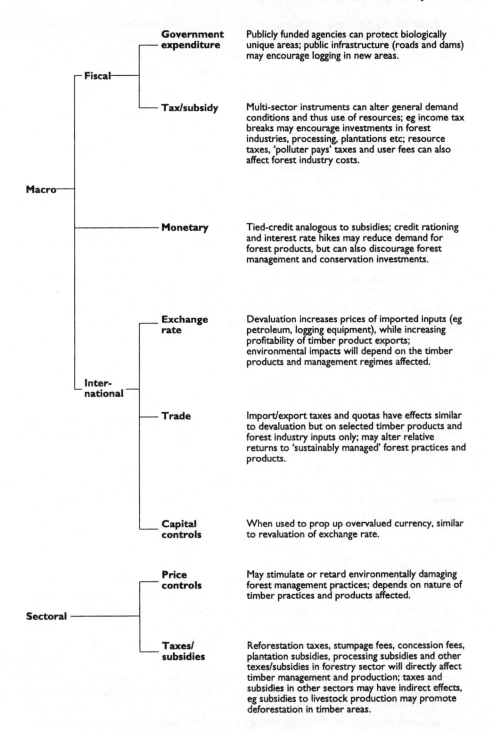

Source Adapted from Bishop, Aylward and Barbier (1991)
Figure 6.1 Economic policy and potential impacts on forest managment

icing payments may also contribute to poor forest management if countries pursue short term policies at the expense of long term sustainable development. More generally, other macro-economic policies (eg fiscal and monetary policy) can affect underlying demand and supply conditions, with knock-on effects in the forestry industry.

The impacts of macro-economic policies on the forestry sector, and subsequently on deforestation, are difficult to analyse. As noted in Chapter 3, the few studies of tropical deforestation often produce conflicting conclusions regarding macro-economic influences on deforestation. For example, in a study of these linkages in tropical forest countries, Capistrano and Kiker (1990) found a negative correlation between debt–service ratios and deforestation. In contrast, Kahn and McDonald (1990) discovered a positive relationships between tropical deforestation and public external debt and annual percentage changes in that debt. On the other hand, the results by Capistrano and Kiker (1990) do show that the ready availability of external funds in the early 1970s in tropical forest countries may have reduced the pressure on the use of domestic forest resources (eg through the need for increased investments and production in forest extraction). Their study also indicates a high correlation between exchange rate devaluation and tropical deforestation. This finding is confirmed by a case study of the links between macro-economic policy and deforestation in the Ivory Coast, which concluded that a devaluation of the exchange rate would stimulate the forest industry (Yao et al 1991). Thus, the indirect impact of debt on increased tropical timber extraction and deforestation, through policy responses to correct or service the debt problem as implemented through structural adjustment programmes, may be significant.

Forest exploitation is directly affected by economic policies aimed specifically at the trade in timber products (eg tax credits or subsidies for forest conversion, afforestation or for wood product exports). The effects of trade policy on forest management and the timber trade is covered in the next chapter. Forest management is also indirectly affected by economic policies which alter incentives and returns in downstream industries or related sectors, such as wood processing, construction and agriculture.

Public investment policy often has important effects on forest-based activities, particularly where transport infrastructure and public services are extended to previously inaccessible forested areas. Such investment may be considered an important subsidy to the logging and wood processing industry, by reducing the costs of gaining access to forest resources, and of bringing forest products to market. Public investment in remote forested areas also acts as an impetus to human population migration and agricultural expansion, frequently associated with forest clearing. The result may be conflicts over land use, or a misallocation of forest resources, as areas designated for timber exploitation are encroached upon by migrant farmers, ranchers, miners and others.

On the other hand, a lack of investment in basic infrastructure and services may impair a country's ability to attain a harvest level that is socially efficient and allows the country to develop its forest industry in line with sustainable practices. The annex to this chapter illustrates how a government policy – eg the requirement that the majority of timber produced by a concession be transformed into sawnwood – can be undermined by poor transport infrastructure and policies that make the export of sawnwood unprofitable for the concession.

The Government of Congo (GoC) has instituted legislation requiring timber concessions to operate a saw mill on site and process 60 per cent of log production. The experience of the CIB concession suggests, however, that given current export conditions, the ability of the concession to conform to this regulation and export sawnwood at a financial profit is being hindered by high transport costs. The latter arise from the poor infrastructure in the Congo for transporting timber products long distances and government-administered transportation charges and taxes. A particularly striking feature of this example is that the company is attempting to operate the concession following sustainable management practices, but current policies and economic conditions make such efforts financially unattractive.

Table 6.4 reveals that, given the need to use a variety of different transportation options, CIB's logging operation could barely break even in 1991. Transport costs accounted for as much as 60–65 per cent of the total fob costs of log production. Meanwhile, the figures for sawnwood indicate that the GoC requirement of domestic processing, combined with the high costs of transport production, made CIB's sawnwood operation unprofitable in 1991. Although transport costs for sawnwood as a proportion of total costs are lower (eg 25–30 per cent) than for logs, the proportion is still substantial. Projections for 1992 suggest that the GoC's decision to raise river transport taxes and freight charges, along with railroad

Table 6.4 Policy failure: Congo case study[a]

	Logs				Sawnwood			
	1991 (actual) Range[b]		1992 (projected) Range[c]		1991 (actual) Range[d]		1992 (projected) Range[d]	
	All figures as % of export price[e]							
Total costs[f]	89.7	101.3	95.1	107.0	99.0	105.4	104.4	113.3
Total transport costs	53.2	64.8	58.6	70.6	24.6	31.1	30.1	30.1
	All figures as % of total costs							
Total transport costs	59.3	64.2	61.6	65.9	24.9	29.5	28.8	34.4
Freight	31.7	40.6	36.4	43.5	14.1	18.1	15.9	21.4
Loading	22.0	18.3	19.5	17.3	6.3	7.0	7.4	7.9
Tax	5.5	5.2	5.8	5.1	2.4	2.3	2.6	2.4
Port fees	–	–	–	–	2.1	2.0	2.9	2.7

Notes [a]Based on fob cost figures for sapelli production and transport from CIB concession in Pokola to port at Point Noire. Sapelli makes up an estimated 80% of annual production.
[b]Low range delivered by government rafts (ATC), high range by private barge (Delacour).
[c]Low range delivered by CIB barge, high range by government barge (ATC).
[d]Low range delivered by CIB in containers, high range by government non-container shipment (ATC).
[e]Average fob export price for 1991 is used as benchmark price for both 1991 and 1992.
[f]In 1991, forest production was halted for 5 months and saw milling for 3 months. During a normal production year costs before transport would be around 6% less.

Source CIB as presented in the chapter Annex

freight fares, does not improve CIB's incentives to continue its logging and saw milling operations on the Pokola concession. Average export unit values would have to rise by approximately 5–15 per cent in 1992 to make sawnwood exports financially profitable. This is extremely unlikely, as current market conditions facing the concession are not favourable.

In sum, the regulation requiring CIB to produce 60 per cent of its timber as sawnwood is not financially realistic, given existing export market conditions and government transport infrastructure, charges and taxes. The regulation may backfire as a strategy to increase government revenues and stimulate value-added processing, unless transportation costs for sawnwood can be drastically reduced through increasing infrastructure investments or reducing taxes and charges. Out of the four timber concessions recently operating in Northern Congo, CIB is the only one currently producing. CIB may also be forced out of the export market, particularly for sawnwood, if current trends in transport costs are not reversed.

CONCLUSION

This chapter has argued that domestic market and policy failures have important implications for forestry management in tropical timber producing countries. However, as many timber products are essentially *tradeable* – goods that may either be exported or serve as import substitutes[10] – the impact of domestic market and policy failures on forest management does not discriminate between timber produced for local or international markets.

If public policies are to be redirected to achieve efficient and sustainable management of tropical timber reserves, then major changes are required. Economic valuation of the impacts of current policies on forest management is essential for determining the appropriate policy responses. Often, however, insufficient economic data and information exist to allow precise estimation of the economic costs arising from domestic market and policy failures. Although in most cases cost estimates as orders of magnitude and indicators of the direction of change are sufficient for policy analysis, in many cases we are not even at this state of 'optimal ignorance' to begin designing appropriate policy responses (Barbier 1994).

Nevertheless, this does not detract from the following conclusions and observations:

- When concessions policy and pricing systems in producing countries provide important incentives to concessionaires to 'mine' the forest, the most practical method for moving towards a global sustainable forest management strategy

---

10 Strictly defined, a *traded* commodity is a good whose production or consumption will affect a country's level of imports or exports at the margins. In addition, a *potentially tradeable* commodity is a good that may not be currently traded but which ought to be if the country adopted optimal trade policies. This latter category would include domestically consumed timber products that benefit from prohibitive tariffs or quotas and for which the marginal cost (at full opportunity cost) of increasing domestic production exceeds the cost of importing.

- such as ITTO's Target 2000 – is to tackle the problem at its source by improving forest sector policy.
- Domestic market and policy failures also have a major influence on the conversion of forest land to agriculture and other uses. As conversion is a prime factor in tropical deforestation, addressing only the domestic market and policy failures that directly affect the forestry sector will by no means be sufficient to halt deforestation and forest degradation in most countries. Other related sectoral policies that affect forest land use decisions may also need to be examined.
- Non-trade related macro-economic policies may also play an indirect role in providing positive or negative incentives for sustainable forest management. As research into economy–environment linkages grows it will be increasingly possible to monitor the incentive role played by these policies.

However, as discussed in Chapter 5, the incentives for sub-optimal use of timber resources may not be linked to domestic and market policy failures alone, but may also stem from distortions in the international timber trade itself. These may arise either through trade restrictions and policies imposed by importing and exporting countries or through the failure of the international timber market to account for the 'global' externalities related to timber production. Having raised the issue of the role of domestic market and policy failures in producer countries in providing incentives for sustainable forest management, the remainder of this book focuses more closely on the issue of the linkages between the tropical timber trade, international policies and the sustainable management of tropical forests.

Annex: Congo – Case Study of CIB[11]

The following case study illustrates how a government policy – eg the requirement that the majority of timber produced by a concession be transformed into sawnwood – can be undermined by poor transport infrastructure and policies that make the export of sawnwood unprofitable for the concession. A particularly striking feature of this example is that the company is attempting to operate the concession following sustainable management practices, but current policies and economic conditions are hindering these efforts.

In north-eastern Congo, near the borders with Cameroon and the Central African Republic, the private foreign-owned company Congolaise Industrielle des Bois (CIB) operates a concession at Pokola of around 480,000 ha. Around 300,000 ha of the concession has been previously logged or is too swampy to use. Logging has occurred over the past 40 years, but with a comparatively small annual volume during the first 20 years. In 1992, the potential timber yield of the remaining 180,000 ha was estimated to be over 2 million m^3 (about 10 to 15 m^3/ha). CIB is attempting to operate this concession on a sustained-yield basis, following recommended management practices based on a 30 year rotation.

Annual production is currently targeted at about 140,000 m^3 of logs, of which 80 per cent is the species sapelli, 10 per cent is *sipo* and the remainder is a mix of ayous (mainly) and other varieties (eg wengue, doussié). However, in 1991 logging was stopped from the middle of March to the middle of July due to poor export demand (the loss of the East European sawnwood market). Total production of logs therefore totalled only 111,639 m^3, with 40,471 m^3 of the logs exported and 71,168 m^3 sawn in Pokola. Of the latter, 33,107 m^3 of sawnwood were exported and 21,613 m^3 were sold for use in the Congo.

A high percentage of the concession's log production is sawn into lumber because of the Government of Congo (GoC) regulation that a timber concession must operate a saw mill on site and that 60 per cent of log production must be processed. However, given current export conditions, the ability of the concession to conform to this regulation and export sawnwood at a financial profit is being hindered by high transport costs. The latter arise from the poor infrastructure in the Congo for transporting timber products long distances and government-administered transportation charges and taxes (see Tables 6.5 and 6.6). Both logs and sawnwood have to be transported 900 km by river and 600 km by rail before they reach Point Noire and can be exported. Given the high cost of the government-run river transport service (ATC), the company has often found it cheaper to use their own or alternative means of transportation. However, CIB cannot avoid government river taxes or port duties, and it must rely on the government-run railway (CFCO) to transport sawnwood and logs to Point Noire.

Table 6.5a shows the actual costs and export values obtained for sapelli exported logs in 1991. Transport costs accounted for as much as 60–65 per cent of the total fob costs of log production. A financial return was possible only if CIB used its own barges or transported logs as rafts. Low river water and high levels of production during the dry season often mean CIB must rely on government or private-run barges as alternative means of transport. As of 1 January 1992, the GoC raised river transport taxes and freight charges, along with railway freight fares. Table 6.5b shows the effects of these cost changes on CIB's returns, using 1991 average export fob unit values. By using rafts and its own barges, CIB might still manage to make some profit on exporting logs, but the margins are reduced significantly. Renting other barges becomes even more unprofitable.

---

11 We are grateful to Henry Stoll and Marc Prevost of Tropical Timber SA, Bâle, and Jean-Marie Mevellec of CIB for assistance in preparing and reviewing the figures presented in this annex.

Table 6.5 Congo – costs and taxes incurred in transporting sapelli logs from CIB's concession at Pokola (CFA per m³; CFA283 = US$1)

Table 6.5a Actual costs, 1991

	Beach costs[a]	River transport costs[b]			Rail Transport costs[c]			Port costs[d]			Total transport costs fob	Total costs fob	1991 average export price fob
		Taxes[e]	Freight	Total	Loading	Freight	Total	Loading	Tax	Total			
ATC barges[f]	21,000	2,503	13,810	16,313	3,102	9,679	12,781	7,541	500	8,041	37,135	58,135	57,553
Delacour barges[g]	21,000	2,503	13,977	16,480	3,102	9,679	12,781	7,541	500	8,041	37,302	58,302	57,553
CIB barges	21,000	2,503	7,700	10,203	3,102	9,679	12,781	7,541	500	8,041	31,025	52,025	57,553
ATC rafts[f]	21,000	2,361	6,686	9,047	3,835	9,679	13,514	7,541	500	8,041	30,602	51,602	57,553
CIB rafts	21,000	2,361	7,700	10,061	3,835	9,679	13,514	7,541	500	8,041	31,616	52,616	57,553

Table 6.5b Projected costs, in 1992 (after price and tax changes)

	Beach costs[a]	River transport costs[b]			Rail Transport costs[c]			Port costs[d]			Total transport costs fob	Total costs fob	1991 average export price fob
		Taxes[e]	Freight	Total	Loading	Freight	Total	Loading	Tax	Total			
ATC barges[f]	21,000	2,656	14,589	17,245	3,102	12,218	15,320	7,541	500	8,041	40,606	61,606	57,553
Delacour barges[g]	21,000	2,656	13,977	16,633	3,102	12,218	15,320	7,541	500	8,041	39,994	60,994	57,553
CIB barges	21,000	2,656	7,700	10,356	3,102	12,218	15,320	7,541	500	8,041	33,717	54,717	57,553
ATC rafts[f]	21,000	2,758	7,235	9,993	3,835	12,218	16,053	7,541	500	8,041	34,087	55,087	57,553
CIB rafts	21,000	2,758	7,700	10,458	3,835	12,218	16,053	7,541	500	8,041	34,552	55,552	57,553

Notes: [a] Costs of harvesting, transporting and loading logs at concession port (Pokola); [b] Pokola to Brazzaville, approximately 900 km; [c] Brazzaville to Pointe Noire, approximately 600 km; [d] Point Noire port; [e] River tax and port duty; [f] GoC river transport company; [g] Private river transport company

Source: CIB

Table 6.6 Congo – costs and taxes incurred in transporting Sapelli sawnwood from CIB's concession at Pokola (CFA per m³; CFA283 = US$1)

Table 6.6a Actual costs, 1991

	Beach costs[a]	River transport costs[b]			Rail Transport costs[c]			Fees	Port costs[d]			Total transport costs fob	Total costs fob	1991 average export price fob
		Taxes[e]	Freight	Total	Loading	Freight	Total		Tax	Total				
Non-container														
ATC[f]	90,000	2,491	13,978	16,469	8,993	9,141	18,134	2,500	500	3,000		37,603	127,603	121,036
Delacour[g]	90,000	2,396	12,094	14,490	9,173	9,141	18,314	2,500	500	3,000		35,804	125,804	121,036
Mossimbi[g]	90,000	2,396	11,900	14,296	9,173	9,141	18,314	2,500	500	3,000		35,610	125,610	121,036
CIB	90,000	2,396	7,700	10,096	9,173	9,141	18,314	2,500	500	3,000		31,410	121,410	121,036
Container														
ATC[f]	90,000	2,491	13,978	16,469	7,979	9,141	17,120	2,500	500	3,000		36,589	126,589	121,036
Delacour[g]	90,000	2,396	12,094	14,490	7,533	9,141	16,674	2,500	500	3,000		34,164	124,164	121,036
Mossimbi[g]	90,000	2,396	11,900	14,296	7,533	9,141	16,674	2,500	500	3,000		33,970	123,970	121,036
CIB	90,000	2,396	7,700	10,096	7,533	9,141	16,674	2,500	500	3,000		29,770	119,770	121,036

Table 6.6b Projected costs, 1992 (after price and tax changes)

	Beach costs[a]	River transport costs[b]			Rail Transport costs[c]			Fees	Port costs[d]			Total transport costs fob	Total costs fob	1991 average export price fob
		Taxes[e]	Freight	Total	Loading	Freight	Total		Tax	Total				
Non-container														
ATC[f]	90,000	2,820	16,919	19,739	10,831	12,389	23,220	3,700	500	4,200		47,159	137,159	121,036
Delacour[g]	90,000	2,725	12,094	14,819	11,011	12,389	23,400	3,700	500	4,200		42,419	132,419	121,036
Mossimbi[g]	90,000	2,725	11,900	14,625	11,011	12,389	23,400	3,700	500	4,200		42,225	132,225	121,036
CIB	90,000	2,725	7,700	10,425	11,011	12,389	23,400	3,700	500	4,200		38,025	128,025	121,036
Container														
ATC[f]	90,000	2,820	16,919	19,739	9,817	12,389	22,206	3,700	500	4,200		46,145	136,145	121,036
Delacour[g]	90,000	2,725	12,094	14,819	9,371	12,389	21,760	3,700	500	4,200		40,779	130,779	121,036
Mossimbi[g]	90,000	2,725	11,900	14,625	9,371	12,389	21,760	3,700	500	4,200		40,585	130,585	121,036
CIB	90,000	2,725	7,700	10,425	9,371	12,389	21,760	3,700	500	4,200		36,385	126,385	121,036

Notes [a] Costs of harvesting, transporting and processing at concession port (Pokola). In 1991, forest production was halted for 5 months and saw milling for 3 months. During a normal production year, beach costs would be around CFA 85,000; [b] Pokola to Brazzaville, approximately 900 km; [c] Brazzaville to Pointe Noire, approximately 600 km; [d] Point Noire port. Fees include fob port costs and taxes; [e] River tax and port duty; [f] GoC river transport company; [g] Private river transport company.

In comparison, Table 6.6a indicates the actual costs and export prices for sapelli sawnwood produced by the CIB concession. Although transport costs as a proportion of total fob costs are lower (eg 25–30 per cent) than for logs, this proportion is still substantial. The result is that exporting sawnwood is unprofitable for the concession. Moreover, the loss is greatest if transported downriver by the government-run service, and the least if CIB transports the sawnwood itself. The higher transport charges and taxes raise the proportionate share of transportation costs (ie to 30–35 per cent) and further increase the losses from sawnwood exports (see Table 6.6). For example, according to Table 6.6b, average export unit values would have to rise by approximately 5–15 per cent in 1992 to make sawnwood exports financially profitable. This is extremely unlikely, as unfavourable market conditions continue to plague the concession. Log production in 1992 was targeted at 145,767 m^3. However, production has had to be suspended indefinitely due to falling demand and poor price trends that have made operations unprofitable, given the high costs involved.

In sum, the regulation requiring CIB to produce 60 per cent of its timber as sawnwood is not financially realistic, given existing export market conditions and government transport infrastructure, charges and taxes. The regulation may backfire as a strategy to increase government revenues and stimulate value-added processing, unless transportation costs for sawnwood can be drastically reduced through increasing infrastructure investments or reducing taxes and charges. Out of the four timber concessions recently operating in northern Congo, CIB is the only one with operations still in place. CIB may also be forced out of the export market, particularly for sawnwood, if current trends in transport costs are not reversed.

7
TRADE INTERVENTIONS IN EXPORTING COUNTRIES

The following chapter examines the use of timber trade export restrictions by *tropical timber exporting countries*, and the implications of such restrictions for deforestation in these countries. The possible effects of liberalizing these export restrictions will also be briefly reviewed. Finally, the chapter will discuss how domestic environmental policies in tropical forest countries can affect their timber trade. Discussion of the import barriers that affect tropical timber exporting countries will be the main focus of Chapter 8.

RESTRICTIONS BY EXPORTING COUNTRIES

In the past, export taxes were used by tropical timber exporting countries primarily as a means of raising revenue from the export of wood in rough. Although there have always been difficulties in administrating and collecting export taxes on logs, such taxes were generally higher than stumpage fees and generated more government revenue (Gillis 1990; Repetto and Gillis 1988; Grut, Gray and Egli 1991).

However, exporting countries have increasingly been imposing export taxes and bans for other strategic purposes. One justification often cited for these policies is that they compensate exporters for the import barriers in developed economy markets by making the price of raw logs higher to the processors in the importing country while reducing the cost disadvantage faced by domestic processors within the timber producing countries. This strategy usually has as its primary aim the creation of more export revenues and employment for the forestry sector; and, for the larger exporters such as Malaysia and Indonesia, the complementary objective of capturing large shares of the international markets for value-added tropical timber products. Some tropical timber exporting countries also claim that their policies will reduce harvesting pressure on the forests by increasing value added per log extracted.

Several authors have recently reviewed the role of export taxes and bans in encouraging forest-based industrialization and sustainable timber management in tropical forest countries.[1] Initially, the preference seems to have been for export

---

1 See, for example: Barbier (1993); Gillis (1990); Vincent and Binkley (1991); and Vincent (1992b). See also other studies cited in Barbier (1994) and in Perez-Garcia and Lippke (1993).

taxes, through employing escalating rates. For most countries, export tax rates on logs ranged between 10 and 20 per cent. Export taxes on sawn timber, veneer and plywood have been negligible. Where sawn timber exports were taxed, rates were typically half that of logs. More recently, the use of export tax structures to promote forest-based industrialization has been largely replaced by log export bans in tropical forest countries, although export taxes are still being used in certain regions and for specific timber products (see Table 7.1).

Tropical timber export taxes and bans have proved only moderately successful in achieving the desired results in SE Asia. For example, although expanded processing capacity was established in Malaysia, the Philippines and Indonesia, it

Table 7.1 Export taxes and bans on tropical timber

Country	Tax rate/Export policy	Remarks
Cameroon	2% of valuer mercurial.	Export taxes are also reported to be 11% of average fob prices.
Central African Republic	· *Logs* US$ 11.45 per m³ for red woods, US$ 11.07 for white woods. *Sawn timber* US$ 250 per m³. *Veneer* US$ 2.90 per m³.	Tax is imposed on processed wood export; roundwood equivalents are much lower.
Côte d'Ivoire	Specific rates variable by species.	*Sapelli* US$ 57.49 per m³. *Sipo* US$ 89.20 per m³. *Assamela* US$ 138.27 per m³.
Ghana	Log export ban enfored since 1979.	
Indonesia	Log export ban since 1985 replaced with export taxes of US$500–4,800 depending on species in May 1992. Since May 1992, specific export tax of US$250 on each m³ of dry and wet veneers.	Specific export taxes on sawn timber introduced in 1989; plywood exempt from all export taxes.
Liberia	Ranges from US$ 1.44 per m3 for low valued species to US$ 58.57 per m³ for high valued species (eg Sipo).	Tax is called the Industrialization Incentive Fee and is imposed on logs only.
Malaysia:		
Peninsular	Log export ban since 1971.	
Sabah	No specific export tax or ban but see remarks.	The Sabah timber royalty has a strong export tax feature: the royalty rate for log exports is almost 10 times the rate for logs used domestically.
Sarawak	15% ad valorem of fob log values.	Tax applies to one hardwood species only.
Papua New Guinea	10% of fob log values.	Tax reported to have been widely evaded through transfer pricing. Log export ban proposed.
Philippines	Log exports restricted to 25% of annual allowable cut since 1979.	Ostensibly to control deforestation.

Source Gillis (1990), updated with data provided by FAO

was achieved at high economic costs, both in terms of the direct costs of subsidization as well as the additional costs of wasteful and inefficient processing operations.[2]

In addition, as a long-term forest-based industrialization strategy, ensuring export sales of processed products through denying processors in other countries access to logs may prove difficult to sustain (Bourke 1988). Importers of SE Asian logs, such as Japan, have been known to substitute other raw materials (eg cement, steel, plastics, fiberboard) for timber, and alternative sources of supply, such as saw logs from temperate and other developing regions (Bourke 1988; Vincent, Brooks and Gandapur 1991; see also Chapter 4).

In Indonesia, the *ad valorem* export tax on logs was doubled from 10 to 20 per cent in 1978, while most sawnwood and all plywood were made exempt. Beginning in 1980, controls on the export of logs were progressively enforced, until an outright ban was introduced in 1985 (Gillis 1988).[3] The export tax structure created effective rates of protection of 222 per cent for plywood manufacture, and the drop in export revenue to the government from diverting log exports was not compensated by any gain in value added in saw milling, resulting in a loss of US$ 15 per m^3 at world prices. The consequence has been the creation of inefficient processing operations and expanded capacity, with implications for the rate of timber extraction and forest management. Gillis (1988) has estimated that over 1979–82, due to the inefficient processing operations resulting from this policy, over US$ 545 mn in potential rents was lost to the Indonesian economy, or an average cost of US$ 136 mn annually. Moreover, as the switch from log to processed exports occurred at a time when forest product prices were falling sharply in real terms, the cost to the economy in export earnings was high. Over 1981–4, the net loss in export earnings amounted to US$ 2.9–3.4 bn or approximately US$ 725–850 mn annually. Additional losses were incurred through selling plywood below production cost, which amounted to US$ 956 mn in 1981–4, or US$ 239 mn annually (Fitzgerald 1986).

Although the switch to value-added processing of timber initially slowed down the rate of timber extraction, the inefficiencies and rapidly expanding capacity of domestic processing may have actually increased the rate of deforestation over the medium and long term.[4] For example, by the early 1980s, the major operational

---

2 For further details see Barbier (1987 and 1993); Constantino (1990); Gillis (1988 and 1990); Repetto and Gillis (1988); Vincent and Binkley (1991); Vincent (1992a and 1992b).

3 In May 1992, due to the threat by the US to take Indonesia to GATT. The export ban was replaced by export taxes of US$500–4,800 per m³ depending on species. This has effectively perpetuated the complete ban on log exports through prohibitive taxation.

4 Barbier (1987) points out that much of the reported decline in log production over the initial period 1979–82 can be attributed to depressed world prices for all timber products and therefore was not necessarily attributable to less exploitation because of increased processing activities. Moreover, there has always been a discrepancy between officially reported harvesting levels and rates based on processing industry output. For example, official logging statistics suggested that total log production

inefficiencies in domestic processing due to high rates of effective protection in Indonesia led to the lowest conversion rates in Asia. As a result, for every m3 of Indonesian plywood produced, 15 per cent more trees had to be cut relative to ply mills elsewhere in Asia that would have processed Indonesian log exports (Gillis 1988). Thus the protection given to Indonesian mills not only increased rather than reduced total log demand, but the gross operational inefficiencies also ensured that millions more logs may have been harvested than if a more efficient policy to boost domestic processing capabilities had been implemented.

More recent analysis of the impacts of the log export ban in Indonesia on the efficiency of timber processing industries confirms that the policy has not increased wood recoveries and thus reduced log consumption compared to pre-ban levels (Constantino 1990). By depressing prices, the log export ban has led to the substitution of wood for other factor inputs, with substantial wood recovery losses in saw milling and a slower growth in wood recovery than other factor productivities in ply milling. In both industries, wood consumption has been the main source of output growth between 1975 and 1987, with efficiency gains contributing very little. However, the log export ban has not affected efficiency in the plywood industry so seriously, which could arise from the much newer capital vintages in ply wood milling, its export orientation that requires higher production standards, and possibly its access to better quality logs.

Despite the problems with the export restriction policy, it is still being aggressively promoted by the Government of Indonesia (GoI) to encourage forest-based industrialization. In Indonesia, the export log ban still remains in place, although since 1991 it has been maintained by prohibitively high tariffs on log exports rather than an outright ban. In October 1989 export taxes on sawn timber were also increased substantially to prohibit exports and shift processing activities to plywood (see Table 7.1).[5] A secondary objective of the policy is to eliminate the marginal mills, leaving only the competitive ones operational, thus improving overall industrial efficiency. The implications of the policy for wood recovery, log demand and thus tropical deforestation in Indonesia is less clear.

As discussed above, many log-producing countries see the use of export taxes and bans as the means to compensate domestic processing industries for import barriers faced in developed economy markets. Bourke (1988) has argued that in the face of such barriers, log-producing countries can maintain some degree of cost-competitiveness by:

- either continuing with existing discriminatory policies against log exports in order to subsidize delivered log prices to domestic processors
- or improving the efficiency of processing operations, especially in plant capacity and conversion rates, in order to reduce costs.

---

peaked, before the ban in 1985, at around 25.3 mn m³ pa and declined initially with the ban. In contrast, estimates of log demand based on processing industry output, capacity and conversion rates indicate that log demand reached 27.3mn m3 pa soon after the ban, and of course continues to increase as capacity and output expands.

5 In May 1992 the GoI also imposed a US$250 per m³ tax on exports of wet and dry veneers.

However, it is clear from the Indonesian case cited above that the use of export taxes and restrictions to encourage processing activities can undermine the efficiency of these activities, with counter-productive consequences for forest management and depletion. Such export trade policy distortions only serve to magnify the negative environmental impacts of domestic market and policy failures in producing countries, as discussed in Chapter 6.

The following section looks more closely at the linkages between these export restrictions imposed in producer countries and their impacts on the tropical forests.

THE IMPACT OF EXPORT RESTRICTIONS

Various partial equilibrium models of forest product markets and timber supply have been developed for countries in SE Asia to examine the implications of policies that restrict log and other timber product exports. The studies for Malaysia did not explicitly examine the implications for tropical forest management of the restrictions, but these can be inferred from the conclusions of the analysis. The studies for Indonesia do discuss how each export policy may impact on tropical deforestation through affecting the government's decision to expand the total area of forest land harvested. Finally, for this book, we have developed our own simulation model of Indonesia's forestry sector to examine explicitly the linkage between timber trade, production and tropical deforestation.[6]

Vincent (1989) employed a simulation model of the timber trade of Malaysia, other SE Asian producers and the major importers from the region to determine the optimal tariffs on intermediate and final goods for log, sawnwood and plywood products. The results of the model indicated that the large export taxes imposed by Malaysia and other SE Asian exporters reduced domestic prices in those countries, leading to losses in producer surplus. More recently, Vincent (1992a) constructed an economic model of saw log and sawnwood production, consumption and trade in peninsular Malaysia during 1973–89 to examine in more detail how the log-export restrictions imposed in 1985 affected the forests products industry in peninsular Malaysia, and whether the restrictions generated net economic benefits. Although the export restrictions seemed to stimulate growth in the processing industries and employment, the economic costs were high. On an average annual basis, peninsular Malaysia lost US$ 6,100 in economic value added, US$ 16,600 in export earnings, and US$ 34,300 in stumpage value for every saw mill job created by the log-export restrictions. No attempt was made in either of the above models to link the impacts of timber trade interventions on timber harvesting levels and tropical deforestation.

As part of the forest sector policy review of Indonesia conducted by the FAO and the GoI, a domestic and international trade model of timber products centred on Indonesia was developed (Constantino 1988a and 1988b; Constantino and

---

6 The model and subsequent analysis appeared initially as Annex 1 in Barbier et al (1993).

Ingram 1990). The simulation model had as its main purpose the determination of supply and demand projections for the Indonesian forestry sector, but it was also used to run policy scenarios on different trade interventions, including the implications of Indonesia's log export restrictions. The impacts of different government harvesting policies were simulated through scenario assumptions concerning the elasticity of the supply of forest land in Indonesia. For example, a small elasticity of supply was interpreted as a deliberate government policy to impose sustained yield constraints; whereas large elasticities reflected the GoI allowing expansion of forest land harvested. Using these different elasticity assumptions, the analysis could then focus on whether in response to each policy scenario the GoI should expand the area of forest land harvested, at the risk of greater tropical timber depletion and deforestation, or whether Indonesia would be better off limiting supply through greater harvest restrictions.

Detailed results for each policy scenario are provided in Constantino (1988a).[7] One of the policy scenarios examined in the study is the effects of Indonesia imposing further export restrictions on sawnwood and plywood, either to divert more production to domestic demand (eg sawnwood) or to assert greater 'market power' in world markets (eg plywood). In the model, export restrictions were simulated by assuming no growth in sawnwood and plywood exports beyond current levels. The analysis of export restriction policies was conducted prior to the imposition of Indonesia's prohibitive sawnwood export taxes in 1989 but included the assumption that the ban on log exports would continue.

According to the policy simulation, export restrictions on sawnwood and plywood result in a loss in international market share for Indonesia, lower foreign exchange earnings, and lower employment, labour income, royalty revenues for logging and economic rent to forest land. On the other hand, the restrictions lead to lower domestic prices thus benefiting Indonesian consumers, higher profits in the processing industries and less forest land harvested. Imposing sustained yield constraints on harvest levels as well would improve matters, as employment would decline by less, even less forest land would be logged and potential government rent capture would not fall as much. The overall conclusion of the analysis is that any sawnwood and plywood export restrictions would reallocate processed products to the domestic market and may alleviate some pressure on the forest resource, but at considerable economic costs in terms of labour incomes, employment, foreign exchange earnings and GoI revenues.

Barbier et al (1994) provide details on the timber trade–deforestation model of Indonesia developed as part of this book. The simulation model employed to examine timber trade and tropical deforestation in Indonesia is a static (single-period), partial equilibrium model of the production, consumption and trade of forest products that is related to the impact of log harvesting on forested area. The model has two components:

- a *simultaneous equation system* determining supply and demand in the logging, sawnwood and plywood sectors of Indonesia; and
- a *recursive relationship* determining tropical deforestation.

---

7   The policy scenarios are also summarized and reviewed in Barbier et al (1994)

Each component was estimated separately over 1968–88, before the 1989 tax rises on Indonesian sawnwood exports. The resulting estimated relationships were linked together in the simulation model using 1988 data. The model was then used to examine the varying impacts on Indonesia's timber markets and tropical forests of the 1989 sawnwood export tax policy and other policy interventions. Full details of all these policy simulations can be found in Barbier et al (1993a).

Table 7.2 displays the results of varying levels of sawnwood export taxes in Indonesia in terms of changes in key price, quantity and deforestation variables as compared to the 1988 base case scenario. The main purpose of the policy simulation was to examine the Indonesian policy of imposing prohibitive taxes on sawnwood exports as a strategy to shift processing activities to plywood, and secondly, to determine what effect (if any) the policy might have on timber-related tropical deforestation in Indonesia.

In the model, the sawnwood export taxes appear to have the effect of reducing export demand for Indonesian sawnwood, thus lowering the export price received and the quantity exported (see Table 7.2). The price effects appear to be outweighed by the quantity effects. Although some sawnwood production is diverted to domestic consumption, it is not sufficient to overcome the fall in export demand. Thus overall sawnwood production falls. The impacts on Indonesia's sawnwood markets are more severe the greater the tax. As indicated in the model, a tax rate of 700 per cent imposed in 1988 would effectively choke off sawnwood

Table 7.2 Indonesia – timber trade and tropical deforestation simulation model: Policy scenario – sawnwood export tax (% change over base case)

Key variables	10% tax	50% tax	100% tax	250% tax	700% tax
Prices (Rp/m^3)[b]					
Log border-equivalent price (unit value)	–0.70%	–3.35%	–6.39%	–14.07%	–29.03%
Sawnwood export price (unit value)	–0.98%	–4.72%	–9.01%	–19.84%	–40.93%
Plywood export price (unit value)	–0.05%	–0.23%	–0.44%	–0.97%	–2.00%
Quantities (000 m^3)					
Log production	–0.16%	–0.77%	–1.47%	–3.23%	–6.66%
Log domestic consumption	–0.15%	–0.74%	–1.42%	–3.12%	–6.44%
Sawnwood production	–0.33%	–1.58%	–3.01%	–6.64%	–13.69%
Sawnwood exports	–2.26%	–10.89%	20.79%	–45.79%	–94.47%
Sawnwood domestic consumption	0.55%	2.66%	5.07%	11.17%	23.06%
Plywood production	0.01%	0.07%	0.13%	0.29%	0.60%
Plywood exports	0.01%	0.03%	0.06%	0.13%	0.26%
Plywood domestic consumption	0.05%	0.25%	0.48%	1.06%	2.18%
Deforestation (km^2)					
Total forest area	0.00%[a]	0.02%	0.03%	0.06%	0.13%
Annual rate of deforestation	–0.3%	–1.60%	–3.06%	–6.73%	–13.88%

Notes [a] A negligible increase over the base case loss of forest cover of 44 sq km
[b] Rp = Indonesia Rupiah
Source Barbier et al (1993a)

export demand. With proposed taxes of US$ 250 to 2,400 per m^3, actual government policy would be represented by the higher range of sawnwood export taxes shown in Table 7.2.

Significantly, the imposition of a tax on sawnwood exports does not appear to instigate a major shift of processing capacity to plywood – at least not in the one-period duration of the simulation model. Only at extremely high rates of taxation does this occur even slightly, and it appears that increased domestic plywood consumption is the most noticeable effect. Evidence would suggest that there are several structural factors limiting diversion of production between Indonesia's sawnwood and plywood industry in the short term, such as the much newer vintages of capital in plywood milling, and its export orientation that requires higher production and possibly access to better quality logs (Constantino 1990). The simulation model therefore confirms that, if a sawnwood export tax is to shift Indonesian processing capacity towards greater plywood expansion, then it will be a long term process. Unfortunately, the economic costs of reduced sawnwood exports and production appear to be severe from the outset.

Only at high rates of taxation does the sawnwood export tax have a modest impact on deforestation. The inability of the policy to increase plywood output in the short term means that the net reduction in sawnwood production translates into less log extraction. At higher rates of taxation, the fall in log production is much larger, thus resulting in a lower rate of deforestation. However, as noted above, the economic costs particularly in terms of lost exchange earnings of a prohibitive sawnwood export tax are severe. Moreover, if the policy is successful over the long run in shifting processing to plywood production, then one would expect log demand, and thus deforestation rates, also to be revived.

In sum, a prohibitive sawnwood export tax in Indonesia appears to be a high cost strategy for shifting processing capacity to plywood production and export, which may only be successful – if at all – over the long term. In addition, the policy does not appear to be an effective approach for reducing timber-related deforestation in Indonesia. Only modest reductions in deforestation may occur, and these will be shortlived if plywood production begins replacing the lost sawnwood output.

TRADE LIBERALIZATION BY EXPORTING COUNTRIES

Very little evidence exists on the environmental effects of timber trade liberalization – ie the removal of the type of export restrictions that have recently been imposed by tropical timber exporting countries. Anderson (1992) provides a theoretical analysis of the general environmental effects of trade liberalization. Some of the results may be relevant to the timber trade.

According to Anderson, a small country exporting a good whose production incurs environmental damage would only gain from trade liberalization if it also ensured that the appropriate domestic policy to correct the environmental problem was in place. In the absence of an appropriate environmental policy, trade liberalization can actually worsen a small country's environment, and the gains in trade from liberalization may not necessarily compensate for this welfare loss. In other words, a timber-exporting country that has high environmental costs associated with its timber exploitation would actually gain from removing its export restrictions only if, as discussed in Chapter 6, it instigated some of the appropriate timber management and other policies necessary to 'internalize' the environmental costs of its forestry operations. The failure to do so will certainly reduce the welfare gains from trade liberalization, in that the additional environmental costs from increased timber product exports through poor management and domestic policies will not be compensated sufficiently by the increase in export earnings.

A study by Boyd, Hyde and Krutilla (1991) investigates some of the possible connections between broad trade liberalization, ie the removal of export restrictions across the major economic sectors, and tropical deforestation in the Philippines. Preliminary findings indicate that trade liberalization would increase deforestation in the Philippines, and that the effects through industrial logging would be particularly significant. Output and exports in the logging sector increase by 6.5 per cent and 28.5 per cent respectively, while investment and employment in logging increase by 2.8 per cent and 13 per cent respectively. The authors note that the wood-based manufacturing sector, as distinct from the rest of manufacturing, is particularly stimulated by trade liberalization as it is highly export oriented and receives no nominal tariff protection.

However, as discussed in Chapter 6, the Philippines has a poor record of *domestic* policy failures affecting timber management and deforestation, including short term concessions, poor regulatory frameworks and inappropriate pricing policies. For example, Boyd, Hyde and Krutilla (1991) note the production restrictions and log export bans in the Philippines cited in Table 7.1, which are ostensibly to control deforestation, but believe these to be ineffectual. The authors believe that well-enforced restrictions would mitigate some of the logging impacts on deforestation associated with trade liberalization. Improvement in all aspects of forestry policy and timber management to control inefficiencies and environmental costs would clearly reduce these impacts further.

DOMESTIC ENVIRONMENTAL POLICY

Improved domestic environmental policies to regulate stand management, develop sustainable management practices and even to protect forest areas can in turn have important implications for timber production and trade.[8] The timber industry is directly affected by concession terms which require reforestation or rehabilitation of logged areas. When such conditions are strictly enforced, they can add significantly to the costs of timber extraction. As argued in Chapter 6, such measures may be necessary to 'internalize' the *user and environmental costs* associated with timber harvesting. In addition, the forestry sector may be constrained by:

- the creation or expansion of public parks and reserves;

---

[8] For a further discussion and overview of these policies, see ITTO (1990).

- by legal protection afforded to certain endangered species of forest plants or animals; or
- by reservation of forested areas for the exclusive use of indigenous populations.

All these public initiatives can effectively reduce both the scale and the profitability of forestry activities, with obvious implications for trade.

As discussed in Chapter 6, from a social perspective the loss of additional timber trade earnings in the short run can be justified, provided that the gains in terms of achieving long-run 'sustainably' managed timber production and improved environmental benefits exceed these costs. Similarly, different forest land use options – whether for timber production, conversion to an alternative use or preservation as a protected area – must be analysed to determine the relative costs and benefits incurred with each option (see Chapter 3). In practice, it is rare for any such analysis to be conducted, either ex ante or ex post, to determine whether a particular environmental regulation or land use option is appropriate. The failure to do so can lead to underestimation of the full production and trade impacts of environmental policies, and can even undermine the objectives of such policies. We illustrate the issues involved by two examples: the establishment of a national park in Cameroon and the trade impacts of sustained-yield harvesting regulations and improved stand management in Indonesia.

The Korup Project is an ongoing programme to promote conservation of the rainforest in Korup National Park in Southwest Province, Cameroon. The area, which includes Korup National Park and the neighbouring Oban Park in Nigeria, contains one of the oldest and most important rainforests in the world. The conservation efforts of the Korup Project are seen to be necessary to forestall the wholesale destruction of the natural habitat in the area in the face of persistent land-use pressures, including timber extraction, within both Cameroon and Nigeria. A social cost-benefits analysis (CBA) of the project was undertaken on behalf of the Government of Cameroon and World Wide Fund for Nature UK to determine whether establishment of the park was a viable economic option compared to other uses of the forest, in particular commercial logging (Ruitenbeek 1989).

The results of the analysis are depicted in Table 7.3. The CBA includes not only the *direct* operating and capital costs of the project but also the *opportunity costs* of lost timber trade earnings (calculated as lost stumpage value) and lost production from the six resettled villages (lost forest use). Against this must be weighed the *direct benefits* of the project in the form of sustained forest use beyond the year 2010 when the forest would otherwise have disappeared, replacement subsistence production of the resettled villagers, tourism, minimum expected genetic value of the forest resources in terms of pharmaceuticals, chemicals, agricultural crop improvements and so on, and environmental functions – watershed protection of fisheries, control of flooding and soil fertility maintenance. Also included are *induced benefits*, agricultural and forestry benefits of the project's development initiatives in the buffer zone. The external trade credit shows a positive benefit to Cameroon of direct external funding of the project. 'Uncaptured genetic value' is a negative adjustment reflecting the fact that Cameroon will be able to capture only 10 per cent of the value of genetic resources preserved in the park through

Table 7.3 Cost-benefit analysis of land use: Korup Project, Cameroon

		Base case result (npv £000, 8% discount rate)
Direct costs of conservation		–11,913
Opportunity costs		–3,326
Lost stumpage value	–706	
Lost forest use	–2,620	
Direct benefits		11,995
Sustained forest use	3,291	
Replaced subsistence production	977	
Tourism	1,360	
Genetic value	481	
Watershed protection of fisheries	3,776	
Control of flood risk	1,578	
Soil fertility maintenance	532	
Induced benefits		4,328
Agricultural productivity	905	
Induced forestry	207	
Induced cash crops	3,216	
Net benefit – project		**1,084**
Adjustments		6,462
External trade credit	7,246	
Uncaptured genetic value	–433	
Uncaptured watershed benefits	–351	
Net benefit – Cameroon		**7,545**

Note npv = net present value
Source Ruitenbeek (1989)

existing licensing structures and institutions. The negative item – 'uncaptured watershed benefits' – indicates that some of the watershed protection benefits of the project will flow to Nigeria and not Cameroon.

Thus the analysis indicates that the Korup Project offers substantial net economic benefits as a land-use option at the project level and to Cameroon as a whole. Although the loss of short-run timber earnings is a substantial cost, it is more than exceeded by the overall benefits of the project.

The section above on trade liberalization discussed two studies in Indonesia that examine the relationship between forestry policy interventions and tropical deforestation – one by Constantino (1988a) and a more recent one conducted by us (Barbier et al 1994). Both studies have included among their policy scenarios the likely impacts of more stringent harvesting restrictions aimed at encouraging sustained-yield log production and conserving tropical forest resource. Examining the effects of such a scenario is important, as several recent analyses have pointed to considerable problems of policy failures in Indonesian forestry management that are contributing to over-harvesting and clear-cutting of permanent production forests, as well as illegal felling of non-production forests.[9]

---

9   See Barbier (1987) and (1993); Burgess (1989); Gray and Hadi (1990); Sedjo (1987); and World Bank (1989).

As discussed above, in the policy simulations conducted by Constantino (1988a), the GoI has the choice of expanding the area of forest land harvested, at the risk of greater tropical timber depletion and deforestation, or limiting supply through greater harvest restrictions (eg sustained-yield regulations). As well as looking at this policy choice in combination with certain trade interventions (eg sawnwood and plywood export restrictions, devaluation, import tax on plywood), Constantino also examined on its own the decision of whether or not to impose conservative harvesting practices to reduce timber-related deforestation.

An interesting result of the analysis is that the economic implications of the policy depended significantly on whether other tropical forest countries also implement this policy. If international competitors follow conservative harvesting practices, then Indonesia should do the same in order to take advantage of greater employment, foreign exchange earnings and rent capture. On the other hand, if competing countries follow expansionary policies, it is not clear what Indonesia ought to do. In the latter situation, harvest restrictions by Indonesia will lead to a loss in international market share and to declines in employment – but also increases in government revenues.

The model of Indonesia's forestry sector developed by Barbier et al (1994) was also used to examine a policy initiative by Indonesia to implement more 'sustainable' management of its remaining production forests. Correcting forest management policy failures would improve the use of Indonesia's remaining tropical forests, and thus reduce any timber-related deforestation, but would also mean higher harvesting costs per m^3 of wood extracted.

Table 7.4 indicates the most likely effects of such a 'sustainable' management policy scenario, which is represented by an increase in harvesting costs. As there is insufficient data to determine the extent to which costs would rise if such a policy were to be implemented in Indonesia, we have arbitrarily chosen a 25 per cent and 50 per cent harvest cost increase for comparison.

A surprising feature of the scenario is that although log prices are affected significantly by the increased harvest costs, any resulting impacts on the rest of Indonesia's forestry sector seem to be somewhat dissipated. There appear to be several factors at work. First, the price-elasticity of log supply for Indonesia is very inelastic in our model (see Barbier et al 1994 for further details). This is not surprising given that Indonesian policy has been to devote log production solely to supplying domestic processing capacity, which of course has expanded considerably over 1968–88. Second, increased log costs are only one component of the total factor costs of Indonesia's processing industries, and have increasingly become the least important component in recent years (Constantino 1990). Finally, Indonesia's sawnwood and plywood exports seem to be the least affected by the increased harvest costs, which would suggest that external demand factors exert an important counteracting influence.

There is reason to believe that the impacts of the sustainable management scenario on reducing timber-related deforestation may be underestimated in Table 7.4. As discussed in Barbier et al (1994), the deforestation equation in the model can only record changes in forest cover due to changes in overall log production rates. It cannot distinguish between *qualitative* changes in the management of log production that may also affect timber-related deforestation more indirectly, such

Table 7.4 Indonesia – timber trade and tropical deforestation simulation model: Policy scenario – sustainable timber management (% change over base case)

Key variables	25% rise in harvest costs	50% rise in harvest costs
Prices (Rp/m^3)		
Log border-equivalent price (unit value)	41.59%	83.06%
Sawnwood export price (unit value)	4.04%	8.09%
Plywood export price (unit value)	2.86%	5.72%
Quantities (000 m^3)		
Log production	–0.94%	–1.87%
Log domestic consumption	–1.37%	–2.73%
Sawnwood production	–1.89%	–3.77%
Sawnwood exports	–1.03%	–2.05%
Sawnwood domestic consumption	–2.28%	–4.55%
Plywood production	–0.87%	–1.73%
Plywood exports	–0.38%	–0.75%
Plywood domestic consumption	–3.12%	–6.24%
Deforestation (km^2)		
Total forest area	0.02%	0.04%
Annual rate of deforestation	–2.28%	–4.23%

Source Barbier et al (1994)

as controlling residual stand damage, improved replanting and reforestation, reduced high-grading of stands, limiting trespass, improving the incentives to control stand abandonment and encroachment, and so forth. These latter factors may be more important in timber-related deforestation than log extraction alone. In addition, some sustainable management techniques, such as utilizing lesser known species and improved harvesting techniques, may actually increase the amount of logs produced from a given timber stand, thus simultaneously limiting harvest costs and reducing the pressure to exploit the remaining tropical forest.

Finally, it is unlikely that the implementation of any sustainable management techniques would occur in a single year, as implied by the simulation scenario. A longer lead time for implementation would mean more time for the Indonesian forest industries to absorb the increased costs, but it would also imply a longer period before the effects on reducing deforestation would be fully felt.

In sum, Indonesia's timber industries may not be badly affected by the higher harvest costs associated with implementing sustainable forestry management policies. Although there would be some reduction in tropical deforestation due to reduced log production, this direct effect may be less significant than the improvement in forest management and protection resulting from qualitative changes in timber stand management practices and ownership.

CONCLUSION

This chapter has examined the use of export restrictions by tropical timber exporting countries and their implications for management of the resource base. Several conclusions emerge:

- The export restrictions, particularly on logs, have stimulated growth and employment in the processing industries of tropical timber exporting countries. For larger exporters, such as Indonesia and Malaysia, the expansion in processing capacity has contributed to their ability to capture a larger share of international markets. This has proved more difficult for smaller exporting countries.
- However, these objectives have clearly been achieved at high economic costs, and have led to problems of over-capacity and inefficiency in processing. Thus many analysts have questioned the economic wisdom of export restriction policies, particularly their extension to cover processed products such as sawnwood.
- Export restrictions have not really been adopted in order to reduce timber-related deforestation in tropical timber exporting countries, although this objective is sometimes also claimed. The reductions in deforestation are not dramatic, and in any case tend to be shortlived. To the extent that export restrictions succeed eventually in stimulating processing capacity and diverting production to domestic consumption, then log demand will recover and pressure on the resource base will resume. Moreover, any resulting processing over-capacity and inefficiencies may further exacerbate this pressure in the medium and long term.
- Essentially, export restrictions only serve to magnify the problems of policy failures concerning unsustainable forest management and poor forestry regulations identified in Chapter 6. In fact, there is some evidence that well-enforced sustainable forest management policies and regulations would mitigate some of the logging impacts on deforestation that may result from the removal of export restrictions.
- Although the enforcement of sustainable forest management policies and regulations may impact on the trade somewhat by increasing the costs of tropical timber harvesting, the impact does not appear to be that significant, as log costs are only one component of the total factor costs of processed timber products. For timber product exports, supply and demand conditions in international markets appear to outweigh the effects of any sustained-yield restrictions on harvesting.
- Moreover, if competing timber exporting countries adopt sustainable forest management policies and regulations in concert, then they may mutually benefit from the strategy. The risks may be greater for one country adopting sustainable management practices unilaterally if its competitors continue to deplete forest resources more heavily.

The main policy implications of this chapter reinforce the conclusions of Chapters 3 and 6. Improved development and enforcement of sustainable forest management practices and regulations are a more appropriate and direct means for tropical timber exporting countries to reduce timber-related deforestation than interventions to restrict timber product exports. To the extent that export interventions do influence logging practices, they only serve to exacerbate the problems of policy failures concerning unsustainable forest management and poor forestry regulations. Where export restrictions lead to over-capacity and inefficiencies in wood industries, logging pressure on the tropical forest resource may actually increase over the medium and long term.

8

TRADE INTERVENTIONS IN IMPORTING COUNTRIES

Import restrictions affecting all types of wood products are widely employed by the governments of both developed and developing countries. Import barriers are typically intended to protect domestic industry from foreign competition. In addition, they are an important source of government revenue, especially in less developed economies. This chapter investigates the use of timber trade restrictions by *tropical timber importing* countries.[1] Existing policies and trends are described. Economic impacts are considered as well as the implications of import restrictions on the development and management of tropical forests. The potential impact of trade liberalization is reviewed. Finally, this chapter examines how other domestic environmental policies in major importer markets can affect the trade in tropical timber, both directly and indirectly.

RESTRICTIONS BY IMPORTING COUNTRIES

Restrictions on trade in tropical timber products imposed by importing countries can be divided into two general categories: *tariff barriers* and *non-tariff barriers*. The former include all types of official levies, duties, surcharges and entry taxes which apply specifically to goods produced abroad. Tariff charges are typically calculated as a fixed proportion of the value of imported goods and the rate applied often varies according to the nature of the product. Non-tariff barriers can assume a bewildering variety of forms. Quantitative restrictions, such as fixed quotas on imports of particular products or from certain countries, directly limit the amount of foreign goods permitted to reach the market. Other less obvious non-tariff barriers may include health, safety and technical standards, anti-dumping and countervailing duty investigations, import licensing schemes or other measures which effectively increase the costs of importing goods. In addition, certain types of government industrial subsidies may be classed as non-tariff barriers to trade, particularly when they are intended to improve the competitive position of domestic firms relative to foreign producers.

Both tariff and non-tariff barriers to trade can have significant economic im-

---

[1] This chapter relies heavily on an up-to-date review of barriers in forest products trade carried out for this book by Bourke (1992a (see also Annex G in Barbier et al 1993)).

pacts. On the one hand they reduce the absolute *volume of trade*, either directly by means of quantitative restrictions or indirectly by increasing the costs of bringing goods to market. At the same time, import restrictions also distort the *pattern of trade*, particularly when barriers differentiate among products or sources, by making certain goods seem artificially cheap or expensive relative to others.

Import barriers effectively reduce both the quantity and the variety of goods supplied in the market while increasing the prices paid by consumers. In almost every case, therefore, consumers are made unambiguously worse off by import barriers. Foreign producers (exporters) lose both sales and profits, because trade barriers mean that they sell less than they could and, in most cases, at lower prices. These losses are only partly compensated by the benefit that import barriers provide to domestic producers, who may gain market share or a significant cost advantage relative to their foreign competitors.

While import barriers can be an effective means of protecting profits and jobs in domestic industries, they therefore impose a high albeit hidden economic cost to domestic consumers as well as to foreign producers. Import barriers can also act as an implicit tax on domestic producers (and exporters) in other industries, not only by increasing the price of *imported* factor inputs directly, but also through their inflationary impact on general price levels throughout the economy.

Tariff Barriers

Bourke (1988) conducted a study of the implications of timber trade barriers for developing countries, focusing on the Asia-Pacific region. In general, trade barriers in major importing countries were found to be negligible for wood in the rough. However, escalating tariff schedules impose more significant barriers to trade in more highly finished products in most markets. In a reassessment of trends in trade barriers, Bourke (1992a) concludes that conditions in most countries are broadly unchanged today.

Tariff rates for the major developed country markets differentiate between imports from GATT member countries, developing countries and other exporters. Base rates are generally those applied to countries accorded most favoured nation (MFN) status, which includes all GATT members and most developing countries. In addition, most developing countries are eligible for a range of special preferences under bilateral and multilateral agreements which provide for reduced tariff rates on exports from certain countries. The most important of these is the UNCTAD Generalized System of Preferences (GSP), which provides for duty-free or special low tariff rates on certain products in some markets. Most preference systems, however, including the GSP, rely on voluntary concessions from importers, who specify the products involved, the countries to which the preferences apply and other limitations, such as quotas or market share restrictions. These additional restrictions can significantly limit the impact of preferential rates. Recent reductions in MFN rates under multilateral trade negotiations have further eroded the advantage provided to developing country exporters under the GSP system.

Tariff rates for imports into developed market economies are around zero for unprocessed products such as logs and rough-sawn timber but higher for more

Table 8.1 Tariff rates for wood and wood products in major importing markets

Industrialized markets	Imports from developing countries		Imports from other developed market economies		Measures of protection post-Tokyo	
	pre-Tokyo	post-Tokyo	pre-Tokyo	post-Tokyo	Nominal rate[a]	Effective rate[b]
EEC						
Wood in rough	0.0	0.0	0.1	0.0	1.0	1.1
Primary products	2.5	1.9	1.0	0.8	1.6	4.0
Secondary products	2.5	1.5	2.2	1.7	10.5	17.9
Japan						
Wood in rough	0.0	0.0	0.0	0.0	2.3	2.3
Primary products	8.2	7.4	0.3	0.2	2.9	8.5
Secondary products	11.1	4.8	9.6	4.3	12.8	24.3
USA						
Wood in rough	0.0	0.0	0.0	0.0	0.0	0.0
Primary products	11.0	5.6	0.8	0.4	0.3	0.0
Secondary products	3.5	1.7	4.7	2.4	7.6	13.7

Developing markets	All imports post-Tokyo
Africa	
Wood in rough	14.4
Primary products	16.2
Secondary products	24.1
Latin America	
Wood in rough	26.2
Primary products	37.6
Secondary products	52.5
Asia	
Wood in rough	34.1
Primary products	57.8
Secondary products	73.1

Notes [a] The nominal rate of protection (NRP) is calculated as the tariff rate multiplied by the ration of the fob price to the landed duty-free price. The NRP measures the extent to which consumers assist domestic producers through payment of higher prices for goods.
[b] The effective rate of protection (ERP) takes account of tariffs on both inputs and final products as well as other subsidies, taxes etc and thus is a more precise measure of the degree of protection afforded to domestic value added. The ERP is computed according to the following formula:

$$E_i = \frac{t_i - \Sigma a_{ij} t_j}{1 - \Sigma a_{ij}}$$

Where: t_i = nominal tariff on final product (i)
t_j = nominal tariff on imputs (j)
a_{ij} = inputs as a proportion of output
$(1-a_i)$ = value added

Source Bourke (1988)

processed goods (Table 8.1). Under the Tokyo Round of GATT negotiations (ie after 1979), average trade weighted tariffs on primary wood products were expected to decline from 2.4 per cent to 1.7 per cent, for primary wood products, and from 7.8 per cent to 5.7 per cent for secondary wood products. Additional reductions since 1986 have further liberalized trade in forest products (see Bourke 1992a for further details).

Import tariffs for wood products are generally much higher in developing country markets, which also practise tariff escalation to discriminate against processed products and to protect domestic industry (Table 8.1). These markets are still relatively small in comparison to the major developed country importers, but they are of growing importance. Note that many of the major tropical timber producers, despite their obvious cost advantage in raw materials, also practise tariff escalation against the import of processed wood products.

The degree of tariff escalation between raw log and processed timber products can have considerable implications for developing country exporters. All else being equal, higher relative tariffs for processed products will discourage the development of value-added wood processing in exporting countries. Timber products particularly affected by tariffs in developed country markets are plywood, some size and species of sawnwood, reconstituted panels and some wood manufactures. For example, Bourke (1988) cites studies which show that Indonesia, Sabah, West Malaysia, the Philippines and Singapore all enjoyed a significant cost advantage in plain plywood production, relative to Japanese processors, ranging from 9 to 39 per cent in 1980 (based on cost, insurance and freight (cif) prices). When the Japanese 20 per cent duty on imported plywood is added, however, this cost advantage drops significantly and actually becomes negative for Indonesia and Singapore. Another study cited by Bourke reports that in 1975 African timber producing countries had a cost advantage over European producers of plywood ranging from 7 to 33 per cent. Import tariffs imposed by the developed countries effectively negate this cost advantage, slowing the pace of industrialization in exporting nations, reducing incomes and employment and perpetuating their dependence on high-volume, low value-added, raw material exports.

However, Bourke also points out that apparent cost advantage in such cases does not always imply a real *comparative advantage*, in economic terms. To illustrate this point he cites a study which compared the costs of producing plain hardwood plywood in Japan in 1980 with costs in a number of other locations in developing countries (Indonesia, Sabah, West Malaysia, Philippines and Singapore). The study revealed that the primary cost advantage for developing country producers can be attributed to a lower cost of raw materials, ie logs. One possible explanation of this difference is that because developing country producers ship finished products with a higher value per unit volume, they incur lower ocean freight costs.

On closer analysis, however, Bourke concludes that the lower cost of logs enjoyed by developing country producers cannot be fully accounted for by differences in ocean freight costs. It appears that this advantage also reflects the many export controls and taxes in developing countries which discriminate against the export of raw logs in order to encourage domestic processing. The apparent cost advantage of developing country plywood producers is considerably reduced when this implicit subsidy on raw materials is taken into consideration. Of course, as discussed in Chapter 7, one reason why developing countries implement restrictions on log exports is to compensate domestic processors for the barriers they face in export markets.

In his recent reassessment of trade barriers for this book, Bourke (1992a) concludes that import tariffs on tropical forest products in most major importing markets are not a significant barrier to trade and indeed are becoming less so. Tariff barriers in the developed countries are relatively low and declining. Most developing country exporters benefit from even lower rates or duty-free entry under the GSP, hence further liberalization of tariff rates under the GATT will not have much effect. On the other hand, tariff barriers in developing country markets remain high. These barriers may be expected to become more significant as the level of trade and consumption of forest products in these countries continues to grow.

Non-Tariff Barriers

Many of the tariff barriers for timber products faced by developing countries are reinforced by additional non-tariff barriers, which can also have a significant impact on markets. Bourke (1988) identifies a wide range of non-tariff barriers that affect forest product trade, including:

- *Quantitative restrictions*, such as combined tariff/quota (TQ) systems which involve higher tariff rates on all imports above a fixed quota ceiling
- *Measures influencing prices*, including variable levies and 'voluntary' price agreements with exporters
- *Health and technical standards*, for example pest and disease controls or end-use restrictions related to structural performance
- *Customs and administrative entry procedures*, which many exporters claim are unnecessarily complicated and time consuming and which add to the cost of importing products, especially for smaller firms
- *Trade agreements*, including trade association agreements or regional trading blocs, which can also distort trade patterns, for example through voluntary arrangements to control volumes and/or prices
- *Ocean freight*, including not only the high cost of transporting timber products to markets but also restrictions on shipping services, for example government requirements that imports arrive on domestic shipping companies or anti-competitive behaviour by shipping firms
- *Other measures* which can include all manner of direct and indirect government subsidies which give some competitive advantage to domestic industry over foreign timber producers, from tax credits for plantation forestry to research and development (R&D) assistance to the processing industry.

It is often difficult to know whether non-tariff barriers are primarily or largely intended to restrict imports or whether they are legitimate restrictions with important functions other than import control. In addition, because they take so many different forms and their effect on prices is usually indirect, the impact of non-tariff barriers on trade in tropical timber products is extremely difficult to analyse.

According to some participants in the trade, many significant non-tariff barriers do not even appear in legislation or official customs regulations, making it even more difficult to assess their real impact. Bourke (1988) discusses non-tariff barriers to trade in tropical forest products in considerable detail, concluding that as a group they probably have a more significant impact than formal tariff barriers. This assessment is re-affirmed in a more recent reassessment carried out as part of the present book (Bourke 1992a).

One major barrier affecting imports of tropical timber is the TQ system, which occurs in both the EEC and Japanese markets. In the EEC, for example, a TQ system limits imports of both hardwood and softwood plywood, builders' woodwork, fibre-building boards and some furniture items, as well as newsprint. Over-quota volumes face tariffs of 10 per cent. Although most developing countries enjoy duty-free access for their products, quotas effectively limit the amount that they can sell in particular markets. The economic impact of such quota systems is uncertain, however, since it is difficult to estimate the extent to which exporters withhold products in order to avoid the higher duty. Anecdotal evidence suggests that some importers respond to quota systems by concentrating their purchasing early in the year in order to ensure dutyfree supplies, thereby creating problems of seasonal oversupply or shortage. Proposals to convert quota systems into explicit tariffs and to reduce rates over time, as part of the current GATT negotiations, have not yet been implemented.

Other non-tariff import barriers of particular significance to tropical hardwood exporters include stringent health and technical standards in many developed country markets, which developing countries often find difficult to meet. For example, recent moves to harmonize EEC building codes and standards have raised concerns among some exporters (mainly of softwoods) that the new regulations will discriminate against foreign suppliers. Prevailing EEC phyto-sanitary regulations also affect softwood (conifer) imports.

Trends in Trade Restrictions

There has been a general trend towards greater liberalization of world trade in recent years, which has also affected forest product imports to the major developed country consumer markets. MFN tariff rates on wood products have declined across the board. As a result, tropical timber exporting developing countries which benefit from duty-free access or lower tariff rates under GSP and other special preference schemes now compete on a more or less even footing with developed country exporters. Many non-tariff barriers, on the other hand, remain in place but their impact on tropical timber trade is much more difficult to assess. Table 8.2 presents a qualitative assessment of recent trends in the incidence of trade barriers affecting forest products. The table shows that since 1985 the most significant change is the increase in non-tariff barriers, generally motivated by environmental concerns.

The tropical timber industry is directly affected by an increasing number of regulations, policies and consumer movements in major importing markets. These involve proposals and actions aimed at restricting trade in tropical timber, products derived from sustainably managed forests, or even halting the trade

Table 8.2 Trends in the incidence of trade barriers affecting forest products[a]

Import restrictions	Direction of movement		
	1960s–79	1979–85	Since 1985
Tariff	Declining	Declining	Declining
Tariff/quota	Increasing	Static	Static
Total prohibition	Increasing	Static	Static/increasing
Conditional prohibition	Increasing	Static	Static/increasing
Quota	Increasing	Static	Static/increasing
Import licence	Increasing	Static/increasing	Static/increasing
Import procedures	–	–	–
Variable levy	–	–	–
Anti-dumping/countervailing duties	Increasing	Increasing	Static
Voluntary export restraints	–	–	–
Price control	–	–	–
Standards	Increasing	Static/declining	Increasing
Government procurement	Increasing	Static/declining	Static/increasing
Marking and packing	Increasing	Static/declining	Static/increasing
Export restrictions	**1960s–79**	**1979–85**	**Since 1985**
Price controls, levies etc	Increasing	Increasing	Increasing
Quotas, prohibitions	Increasing	Increasing	Increasing

Notes [a]Subjective assessment by I. J Bourke (1992a) based on a review of available information. The assessment does not involve any weighting by the volume of trade.
– = Little or no importance

altogether. One major difference between these proposals and most existing restrictions on trade is that the former are promoted for environmental or conservation reasons, rather than for the usual reasons of protecting or encouraging domestic industry, saving foreign exchange, raising incomes and so on. In many cases the initiative for the new proposals comes from non-governmental organizations (NGOs) rather than from central government or industry groups. The types of measures and actions being promoted fall into three broad categories (Bourke 1992b):

- *Consumer/trade boycotts* These aim to discourage or limit the use of tropical timber or to limit trade to those producing countries which follow recognized standards of forest management. They are typically proposed by consumers, trade associations or, in some cases, local or regional governments through building and construction codes.
- *Direct control or bans* These consist of proposals for national governments to implement tariffs, quotas or other official trade controls which would reduce or eliminate trade in tropical timber or, as above, limit the trade to countries which conform to specified guidelines for forest management.
- *Revenue-generating actions* These are proposals that importing country gov-

ernments or private trade federations should raise new funds to assist tropical timber producing countries in the transition to sustainable management. The proposals include compulsory and voluntary surcharges on timber imports (tropical only or all timber) as well as contributions assessed in other ways.

Informal barriers such as consumer or trade boycotts have attracted a great deal of attention, although their impact to date has been relatively small. Such actions could conceivably have a significant impact, however, if adopted more broadly. Assessing the impact of these measures is complicated by the fact that they have coincided with a period of economic recession in most major importing countries, during which demand for timber products is depressed generally. As reported in Chapter 4, surveys in some developed countries suggest that consumers may be willing to support these actions through higher prices for sustainably produced tropical timber products, but the evidence is mixed.

Proposals for new or higher tariffs, quotas, bans or other direct controls on trade in tropical timber have had little impact so far. Such actions are unlikely to be implemented by consumer governments, due to GATT commitments and the fear of trade retaliation. However, threats of such action, even if never implemented, can also be an effective deterrent to trade. Non-tariff barriers, such as labelling of tropical timber products, are far more likely to be implemented than new tariffs, precisely because such measures fall outside the jurisdiction of the GATT.

Schemes to generate revenue to assist the shift to sustainable forest management have also been widely mooted but to date none has been implemented (Bourke 1992b). One recent proposal put forward by the timber trade associations of the UK and the Netherlands involved a voluntary surcharge on tropical timber imports into the EEC, with revenues to be used to assist producing countries improve their forest management practices. The proposal was subsequently dropped under pressure from producer governments, which felt that it would be discriminatory.

To summarize the discussion in this section, tariff barriers on tropical timber products in the major developed consumer markets are relatively low and declining, reflecting recent government commitments to trade liberalization under the GATT and other bilateral arrangements. Tariff barriers in the smaller developing country markets generally remain much higher, although these also appear to be declining. Non-tariff barriers to the import of tropical timber products, on the other hand, appear to be increasing in significance. In some cases this may reflect backsliding on previous commitments to reduce tariffs, in order to maintain some degree of protection for domestic industry. In many developed countries, however, these trends also appear to reflect growing pressure from media and advocacy groups to reduce consumption of tropical timber which, rightly or wrongly, is perceived to be an environmentally 'unsound' product.

THE IMPACT OF IMPORT RESTRICTIONS

It is extremely difficult to assess the environmental effects of import barriers on tropical forests. This section attempts to trace the linkages between import restrictions affecting the trade in tropical timber products and the management of tropical forests. Through their effects on trade in tropical timber products, import restrictions may be expected to influence the supply of tropical timber, the distribution of value added, forest sector employment and the efficiency of processing. These factors will in turn influence the quality of timber stand management in production forest, as well as the long-term allocation of forest land between the timber industry and alternative land use options.

As discussed above, trade barriers clearly reduce demand for certain types of tropical timber products in importing countries. Hence import restrictions may seem to be an effective means of reducing the pace of logging and deforestation in the tropics. However, the long-run effects of trade barriers on the supply of tropical timber and the management of tropical forests are not so obvious; nor are they clearly beneficial.

While logs and rough-sawn wood may be imported to most major markets duty free and with little other hindrance, more highly processed goods often face significant tariff and non-tariff barriers. The immediate (short-run) effect of these restrictions is to depress prices for processed goods received by developing country exporters. Lower export prices in turn discourage the development of value-added processing in timber-producing countries, preventing an efficient forest-based industrialization.

According to this argument, low demand and prices for processed products from the tropical countries will, in the long run, feed through to reduced demand and lower prices for the main raw material input, namely saw logs. By putting downward pressure on raw log prices, import barriers thus further exacerbate the disparity between stumpage prices and the real economic scarcity value of timber. As discussed in preceding chapters, low log prices will delay the transition of the forest sector from mining of old-growth stocks to efficient management of secondary-growth forests and retard the coordination of processing capacity with timber stocks.[2] Moreover, if the economic return to forestry is depressed by trade restrictions, then alternative uses of forest land, such as plantation agriculture or ranching, will seem relatively more attractive by comparison. Suppressing the timber industry could therefore lead to the permanent conversion of even larger areas of tropical forest land to more environmentally destructive uses.

Of course, there are complications with such an argument. As pointed out by Bourke (1988), although developing countries as a group have a comparative advantage in simply worked wood and wood manufactures, and possibly veneer and plywood, it is not certain that the timber-producing developing countries within this group would necessarily have a comparative advantage in producing all these products. For example, some newly industrialized countries (NICs) that have little or no production forest to speak of, such as Taiwan, South Korea and Singapore, currently have a greater comparative advantage in more advanced processed products, such as plywood and veneer. In the short run, therefore, existing import barriers may discriminate more against the NICs than against the

---

2   For further discussion of the role of stumpage value in forest-based industrialization see Vincent and Binkley (1991).

timber-producing countries. The long-run impact on log prices, however, still holds.[3]

The potential impact of the new environmental initiatives is similar to that of traditional import barriers. Total trade with the largest and most profitable markets would be reduced, forcing producers to cut back production and look for alternative markets. In the short term the rate of tropical deforestation might be reduced but the long-term effects could be perverse. If new markets cannot be developed at home or abroad, the income received by tropical timber exporters would presumably fall, along with the economic incentive to invest in forest management (see Chapter 3). The influence of developed country consumers over tropical forest management would also be reduced by their actions, as the relative importance of these markets declines.

Restrictions on the import of tropical timber products to major consumer markets *can* have an impact on tropical forest management. However, the link is complex and involves both first-order and second-order effects. While import restrictions certainly reduce demand for tropical timber somewhat, it is by no means obvious that they are effective in reducing the pace of logging and deforestation in the tropics. As described in the preceding section, the widespread use of export restrictions by tropical timber-producing countries creates additional distortions in patterns of trade and forest management incentives. These overlapping effects add to the difficulty of isolating the specific impact of import barriers on timber trade flows and forest degradation.

TRADE LIBERALIZATION BY IMPORTING COUNTRIES

Little empirical analysis has been carried out on the effects of removing timber trade import barriers on deforestation or forest degradation. However, a number of studies have attempted to estimate the effects of trade liberalization on both the absolute levels and patterns of trade in forest products. From this information it is possible to infer the impact of trade liberalization on the supply of tropical timber and thus on tropical deforestation.

One of the more comprehensive analyses of the effects of trade liberalization simulated the removal of import tariffs on *wood and wood products* in ten major developed market economies (reported in Bourke 1988). The effects on the imports from developing and developed market economies are indicated in Table 8.3. Two effects are shown. Tariff reduction would lead to *trade creation* by boosting demand through a lower import price. Tariff reduction would also result in *trade diversion*, as relative price changes lead importers to shift some of their purchases to other suppliers.

The analysis suggests that the removal of import tariffs alone (based on official post-Tokyo round MFN rates) would create around US$ 150 mn in additional

---

[3] One difficulty in assessing the importance of this effect is distinguishing the impact of import barriers, which may depress log prices, from the influence of log export restrictions imposed by producer countries, which make tropical logs artificially scarce and thus drive up prices on international markets.

Table 8.3 Estimated trade effects from the removal of post-Tokyo round tariffs in the main developed market economy markets[a]

	Imports from developing economies			Imports from developed market economies		
	Trade creation	Trade diversion	Total effects	Trade creation	Trade diversion	Total effects
Importers	(US$ mn)		(%)	(US$ mn)		(%)
EEC	45.2	20.4	4.0	78.4	−38.6	1.3
Japan	49.2	−0.2	2.7	15.4	0.5	0.8
USA	46.1	−4.5	6.2	56.2	4.5	3.0
All[b]	150.6	−2.0	3.35	731.4	−5.6	8.1

Notes [a]Increase over 1976 trade levels. Trade diversion estimate is an average of low and high estimates. Total effects is an average of low and high estimates of the % of actual imports.
[b]Total of ten developed market economies; others not shown are Austria, Canada, Finland, New Zealand, Norway, Sweden and Switzerland.
Source Bourke (1988)

exports for developing countries – a 3.3 per cent increase – with trade diversion away from developing country exporters amounting to only US$ 1 to 3 mn. For these countries the additional trade created far outweighs any trade diversion. Almost all new trade for the developing countries would involve the EEC (US$ 45 mn), Japan (US$ 49 mn) and the United States (US$ 46 mn). For developed market economies as a whole, trade creation in this simulation totals about US$ 730 mn – an 8 per cent increase – with trade diversion no greater than US$ 12 mn. The bulk of trade creation benefits would accrue to Austria, Canada and Switzerland while the EEC would bear the brunt of negative trade diversion effects (which would still not exceed the Community's gains from trade creation).

Another study reported in Bourke (1988) estimated the combined effect of removing both MFN tariffs and quantitative non-tariff barriers to imports of *wood and paper products* from the developing countries only. The analysis showed that imports of these products would increase over 1980 levels by US$ 638 mn for the EEC, by US$ 40 mn for the USA and Canada and by US$ 10 mn for Japan. Although the estimated gains from trade liberalization would be less for forest products than for other product groupings (such as textiles, clothing, food and beverages), nevertheless the potential increase is significant (about 18 per cent above current levels for all three markets). The largest impact would be felt in the EEC, reflecting both a high level of wood and paper products trade as well as greater existing import barriers in that market.

Further estimates of the gains from such preferential trade liberalization were carried out at a more disaggregated level (Bourke 1988). These revealed potential trade creation for the developing countries ranging from US$ 29 mn for non-coniferous sawnwood (a 0.6 per cent increase over 1980 levels) to US$ 96 mn for plywood exports (an 11 per cent increase) and as much as US$ 1,000 mn for furniture (an estimated 60 per cent increase over 1980 levels of trade). The higher estimated gains for plywood and especially for furniture clearly show the impact of escalating tariff discrimination against processed products.

Although substantial trade creation could result from the elimination of all trade barriers on timber products, it is unlikely that a complete removal of tariff and non-tariff barriers would occur. The effects of a piecemeal or gradual reduction of trade barriers are more difficult to determine. The impact on forest resources and the pace of deforestation is even harder to predict and, as discussed above, will involve both first-order and second-order effects.

All of the studies described above appear to show that liberalization of import barriers to trade in forest products would stimulate demand for exports from the developing countries, with processed product exports benefiting the most. These results suggest that total forest sector output in developing countries would increase and that the process of industrialization would be accelerated. In principle, forest product exporters would enjoy higher prices which would in turn feed back into greater incentives for sustainable forest management. However, in the light of the serious domestic market and policy failures which prevail in the forest sector of most developing countries, an increase in forest product prices and sector output may not lead to greater investment in timber stand management but instead simply exacerbate problems of excessive forest degradation.

Moreover, it is not at all clear from these simulations whether the gains from increased processed product exports would necessarily accrue to tropical forest countries. As discussed in the preceding section, other developing countries with little or no forest resources may enjoy a competitive advantage in timber processing. An increase in world trade in forest products could therefore stimulate logging without significantly improving the pace of forest-based industrialization in tropical forest countries. However, if the tropical forest countries maintain their existing restrictions on unprocessed wood exports, they might have a better chance of capturing the benefits from liberalization of import barriers. Whether this would lead in fact to more efficient use of wood and more sustainable management of production forests remains in doubt.

IMPACTS OF ENVIRONMENTAL REGULATION BY IMPORTING COUNTRIES

In addition to policies explicitly directed at trade, other domestic environmental regulations in timber importing countries can have important implications for the international trade in tropical timber. The following section examines the impacts of domestic environmental regulations, policies and agreements affecting the forest sector, focusing in particular on the impacts of domestic environmental regulations on the international trade in tropical timber.

The tropical timber industry is also affected by environmental regulations aimed at the *temperate* forest industry. For example, recent studies carried out in the United States have suggested that the combination of environmental and trade restrictions on logging in the Pacific North-west may have significant impacts on timber flows and prices both domestically and abroad (Flora and McGinnis 1991; Perez-Garcia 1991). Three harvesting restrictions in the region (mainly affecting Douglas fir) are being implemented almost simultaneously:

Table 8.4 Domestic and export effects of timber trade and environmental restrictions: US Pacific North-west[a]

Table 8.4a Short run

	Base level (1990)	Replanning (all lands) and embargo	Spotted owl (public lands), replanning (all lands) and embargo	Spotted owl (all lands), replanning (all lands) and embargo
		(Percentage change from base level)		
All-owner harvest	11.54[b]	−5	−8	−16
Log export volume	2.92[b]	−12	−16	−25
Lumber shipments	7.76[b]	−2	−3	−5
Lumber export volume	1.76[b]	−3	−7	−16
Domestic log prices[c]	293[d]	+30	+52	+135
Export log prices	450[d]	+48	+66	+137
Domestic lumber prices	309[d]	+9	+14	+31
Export lumber prices	336[d]	+24	+43	+116

Table 8.4b Long run

	Base level (1990)	Replanning (all lands) and embargo	Spotted owl (public lands), replanning (all lands) and embargo	Spotted owl (all lands), replanning (all lands) and embargo
		(Percentage change from base level)		
All-owner harvest	11.54[b]	−4	−6	−12
Log export volume	2.92[b]	−11	−14	−23
Lumber shipments	7.76[b]	−1	−2	−3
Lumber export volume	1.76[b]	−3	−8	−21
Domestic log prices[c]	293[d]	+16	+26	+60
Export log prices	450[d]	+27	+37	+68
Domestic lumber prices	309[d]	+5	+8	+17
Export lumber prices	336[d]	+19	+36	+101

Notes [a]Short run: all policy changes occur in 1990 and their impacts are experienced in that year; long run: three-year lagged impacts
[b]Billions of board feet
[c]Average price of number 2 softwood logs
[d]US$ per 000 board feet
Source Flora and McGinnis (1991)

- *State-log export ban* The 1990 Forest Resources Conservation and Shortage Relief Act passed by the US Congress banned the exports of most state-sold logs, with the exception of Alaska and 25 per cent of state-owned timber in Washington.
- *Replanning* Coinciding with a revised outlook for harvests on private lands, replanning of timber harvests on National Forest lands will cause a substantial reduction in regional harvests.

■ *Spotted owl reservations* Specific geographic zones in the region, mostly on public old-growth forest lands, have been proposed as reserved habitat for the spotted owl. In addition, revised harvest scheduling has been proposed for all forest lands, private and public, considered essential to restoring owl populations.

The spotted owl reservations are a clear example of an environmental restriction on timber operations in the region; replanning has elements of both a timber and environmental restriction, whereas the log export ban is unambiguously a trade restriction. Flora and McGinnis (1991) have analysed the incremental and cumulative impacts of these restrictions on domestic and export timber flows and prices for both logs and milled lumber from the Pacific North-west. Their results are shown in Table 8.4 and indicate substantial impacts on domestic and export trade from environmental restrictions both on their own and in combination with trade restrictions.

In the short run, the decline in domestic harvesting, lumber (sawnwood) shipments and log and lumber exports will be significant, with the addition of spotted owl reservations at least doubling the impacts. The greatest impact of the reservations will be on sawnwood export volumes. Even greater effects will occur on domestic log prices and export log and lumber prices. Replanning and the export embargo are expected to raise these prices by around 25–50 per cent, whereas the addition of the spotted owl reservations on public lands further raises prices by around one half, and imposing them on all lands effectively doubles these prices. Domestic sawnwood prices are affected similarly, but the impacts are less severe.

The results could include an export premium for Pacific North-west logs, increased revenues for log sellers in the region, but a cost-price squeeze on US and, to some extent, foreign lumber producers. Moreover, as timber harvests decline, both log and sawnwood exports fall, with a disproportionate drop in log exports. Although the export premium will mean that more logs and lumber are exported, foreign importers will respond to higher prices and lower supplies by purchasing both logs and lumber from elsewhere. Thus, over the longer term, the price effects are projected to subside somewhat as both domestic and foreign timber users find ways to substitute for Pacific North-west supplies. As a consequence, the long-run volume effects on harvesting, domestic supplies and exports will also be less severe.

The wider domestic and global impacts of the trade and environmental restrictions on logging in the Pacific North-west have been examined by Perez-Garcia (1991). This study used the CINTRAFOR Global Trade Model to assess the effects of environmental legislation in terms of reductions of the volume of timber produced from public and possibly private lands, amounting to as much as 44 per cent initially.[4] It also examined the 75 per cent log export ban for Washington state lands, which will increase the domestic supply of logs. The results of the study indicate significant global impacts of the environmental restrictions, both

---

4   See Perez-Garcia and Lippke (1993) for further details, or Annex K in Barbier et al (1993).

on their own and in conjunction with the trade restriction. The projected impacts of the environmental restrictions on their own are:

- An increase in saw log prices around the globe both in the short and long term, led by large increases in the Pacific North-west region.
- Global saw log production decreases by 0.4 mn m3 in the short run and 1.2 mn m^3 in the long run; however, large falls in the Pacific North-west and neighbouring regions are somewhat offset by increased production in other US regions, Canada, Japan, Sweden, Finland, Western Europe, Chile and New Zealand.
- The net short- and long-term effects of a decline in saw log production on log consumption by mills is to decrease consumption in both the US west and inland regions, as well as in major importing countries such as Japan and South Korea. In contrast, more logs are consumed in the US south and Canada.
- In the sawnwood and plywood markets, prices also increase as a result of reduced global supplies in the short and long term. Over the long term, production of sawnwood and plywood decreases in the US West, whereas Canada increases sawnwood exports and the US South increases both sawnwood and plywood exports. Japan is forced to increase imports to replace falls in domestic lumber production.

The effects of combining the environmental restrictions with the 75 per cent export ban for Washington state forests offsets some of the above impacts in the short run for the US west, but has cumulative effects on other markets. In the long run (eg the year 2000), many of the effects of the trade restriction are overshadowed by the impacts of the environmental restrictions.

However, if other log producers such as Russia or tropical forest countries substantially increase their production and exports to compensate for the restrictions in the Pacific North-west, the impact of the US log shortfall on international markets, prices and timber harvesting is reduced. The market share of the Pacific North-west will not only fall as log costs for saw mills and plywood mills increase, but log importing countries will also experience an increase in their log costs. Increased substitution from other timber sources and of non-wood products is expected, as is a general increase in consumer costs globally (Perez-Garcia 1991). On a global scale, the environmental implications of increasing timber harvests in other regions of the world may offset the environmental gains associated with the environmental restrictions in the Pacific North-west. This will particularly be the case if the expansion of harvesting in the Soviet East and tropical countries proceeds without correction of the domestic policy and market failures and trade distortions outlined in Chapter 6.

CONCLUSIONS

This chapter has examined barriers to the import of tropical timber in major consumer markets, including traditional protectionist measures as well as more recent trends in environmentally motivated restrictions. The economic and environmental effects of import barriers were explored, with a focus on the potential impact of liberalization on trade in tropical timber products and tropical deforestation. The chapter also examined other domestic environmental policies in major consumer countries which may affect trade in tropical timber and tropical forest management. The major conclusions are:

- Import tariffs on tropical forest products are generally low and declining in the major developed consumer markets. Tariff barriers on wood in the rough are virtually non-existent, while low positive tariff rates apply to some processed products. Tariffs are higher for all products in most developing country markets, but the latter do not currently import significant volumes of tropical timber.
- Tariffs do not appear to be a significant barrier to trade and are becoming less so. Moreover, most developing country exporters benefit from low rates or duty-free entry under existing preference systems, so further liberalization would have little effect on trade.
- Non-tariff barriers to trade in tropical forest products also discriminate against processed wood imports. These barriers are extremely diverse and probably have a more significant impact than formal tariff barriers. Non-tariff barriers appear to be on the increase in most markets, many of them motivated by environmental concerns.
- The combined effect of tariff and non-tariff barriers on imports of processed tropical timber products into major consumer markets can hinder the ability of producer countries to develop their own value-added wood-processing industry.
- However, the negative impact of import barriers on the pace of forest-based industrialization in developing countries is offset by the many restrictions on raw wood exports imposed by most tropical timber exporting nations.
- Although import barriers in developed market economies are reducing trade flows from developing countries somewhat, the environmental implications of these policies are difficult to determine. In the short run, removal of import barriers in developed market economies would increase demand, raise producer prices and stimulate forest sector output in exporting countries. In light of the market and policy failures which discourage sustainable forest management in many tropical forest countries, an increase in sector output could accelerate the pace of deforestation and land degradation.
- In the long run, increased demand for forest products will raise stumpage values, which may increase the economic incentives to invest in more efficient wood processing and in better management of production forests in tropical timber producing countries. This could eventually help to reduce over-exploitation of old growth forests, provided that the appropriate domestic forest policies are also implemented.
- Recent proposals to restrict trade in tropical timber for environmental reasons have arisen in some of the major consumer countries. The short-term impact may be to decrease the pace of tropical deforestation but in the long term developing country producers would be worse off and incentives to invest in sustainable forest management may be reduced.

- Other domestic environmental regulations in timber importing countries can have important implications for the international trade in tropical timber. In particular, regulations which increase costs to temperate forest industry or decrease the available inventory may lead to higher world prices, stimulating supply from other areas. Given the prevailing market and policy failures prevalent in many tropical forest countries, this could mean that more stringent environmental regulation in temperate zones will, indirectly, accelerate deforestation in the tropics.

9

MEASURES FOR THE SUSTAINABLE USE OF FORESTS

The following chapter identifies several trade-related policy options which have been proposed as a means to encourage more sustainable management of tropical timber production forests. Some of these policies involve interventions in the tropical timber trade, with varying distorting effects. As these interventions could be implemented either unilaterally or multilaterally, we discuss the general feasibility of both approaches. A more detailed assessment of each policy option is conducted in Chapter 10.

IDENTIFYING AND ASSESSING OPTIONS

There is a wide range of policy options for the tropical timber trade: from no intervention, to a complete trade ban, to trade liberalization. There are, of course, also numerous policy options lying between these extremes. Despite their diversity, in this book we refer to these policy options broadly as *trade policy options*. However, some are not strictly trade related but do affect timber production and trade indirectly by influencing the incentives for sustainable timber management. In Chapter 10 we discuss in more detail whether policies that are trade interventions in a narrow sense are more appropriate than other policy approaches. A comprehensive list and explanation of the policy options identified are provided below.

Do Nothing

The 'do nothing' scenario implies that existing tropical timber related trade policies remain as they are today and that no additional policy interventions to affect trade patterns and/or the incentives for sustainable forest management are implemented.

Measures to Alter the Pattern of Trade

Policies may be specifically designed with the objective of changing the patterns of tropical timber trade flows. There are several policy options that could achieve this objective:

- *Trade liberalization* This involves the removal of all existing international tropical timber trade interventions by importers, exporters, or both. Trade liberalization has been discussed in Chapters 7 and 8.
- *Tropical timber trade ban* This is a 'command and control' regulatory instrument whereby compliance is mandatory and some form of monitoring and enforcement is required. A trade ban may take several different forms. For example, a *complete* ban on all international trade in tropical timber products would effectively end all legal trade. A *selective* ban on all 'unsustainably produced' tropical timber would make all trade in tropical timber products illegal if they were not produced in compliance with sustainable forest management objectives. However, trade in tropical timber products would still be legal if they were produced according to sustainable forest management criteria.
- *Quantitative restrictions* These are essentially restrictions, or quotas, on the quantity of international trade in tropical timber products. Quantitative restrictions are generally achieved through *direct control on the quantity of trade*. This form of quantitative restriction is another example of a 'command and control' regulatory instrument, whereby compliance is mandatory and some form of monitoring and enforcement is required. A 100 per cent restriction on the quantity of trade is effectively the same as a complete ban on all international trade in tropical timber. Trade may also be limited by a certain proportion, for example trade in tropical timber products may be reduced to 80 per cent of current levels by a direct regulation system. The trade limits may be targeted to specific product categories (eg timber products derived from endangered species) or specific sources (eg a quota conditional on the level of sustainable offtake). For a product differentiating quantitative restriction to be implemented successfully in practice a system of product identification and certification would be required.
- *Trade taxes* These increase the costs associated with trading timber products, which in turn discourage producers and consumers from trading in them. In order for a tax to be effective in achieving a reduction in the quantity of trade it must stimulate producers or consumers to change their patterns of resource use. In this way, the tax is an 'economic instrument' that affects the economic incentives for international trade in tropical timber. The tax may be implemented either at the exporting or the importing end of the international tropical timber trade. However, where the tax is implemented will affect the extent to which the cost of the tax is passed on to the producers or the consumers. A tax may be also employed on differentiated products that enter the international trade. For example, a tax could be imposed on unsustainably produced tropical timber in order to discourage the trade in such products. For a product differentiating tax to be implemented successfully in practice a system of product identification and certification would be required.
- *Trade subsidies* These are also an 'economic instrument' that stimulates producers and consumers to change their patterns of resource use. However, unlike a tax, a subsidy essentially reduces the costs associated with the trade and encourages higher levels of trade in timber products. Employing this economic instrument across the tropical timber trade in general would not,

therefore, reducing the quantity of international trade. However, a subsidy could be used on differentiated products that are traded internationally to alter patterns of trade. For example, a subsidy could be levied on sustainably produced tropical timber in order to encourage the trade in this timber rather than the trade in timber from unsustainably managed tropical forests. For a product differentiating subsidy to be implemented successfully in practice a system of product identification and certification would be required.

Measures to Raise Revenue for Sustainable Management

A redistribution of revenue raised by the trade may be used to implement sustainable forest management in tropical countries. There are several ways in which the revenue could be raised:

- *Redirection of existing revenue from the trade* Existing revenue that is captured by government intervention in the trade in timber products may be redirected and targeted for encouraging sustainable forest management in tropical countries. For example, the revenue raised by value added tax (VAT) on imported products, import duties or export levies on tropical timber products could be channelled through a multilateral or bilateral funding mechanism to encourage sustainable forest management.
- *Appropriation of additional revenue from the trade* Rather than relying on the redirection of existing revenue captured by governments from the trade *additional funds* may also be raised from the trade. A *trade surcharge* could be implemented as a levy on the trade in tropical timber products in order to raise revenue. This could be implemented either by the importing or exporting countries. Again, where the tax is implemented will affect the extent to which the cost of the tax is passed on to the producers or the consumers. The trade surcharge could be either voluntary or mandatory.
- *Additional funding external to the trade* New development assistance funding could be made available to tropical forest countries from existing or future bilateral and multilateral aid flows. If necessary, additional funds could be raised through a wide range of levies or mechanisms – not necessarily related to the tropical timber trade – for sustainable tropical forest management.

Certification

Certification is a means of providing information about the traded timber product and is not strictly a policy option. It can be used to inform consumers that timber comes from sustainably managed tropical forests. However, certification may be considered an important prerequisite for implementing many of the policy options described above. For example for a sustainably produced timber product differentiating tax or subsidy to be successfully implemented, identification and certification of timber products from sustainably produced tropical forest is required. Certification is also required if the policy option is to ban unsustainably produced tropical timber. There are many ways of establishing a certification system – for example, on a producer country level, on a concession level, or at the product level. The mechanisms for certifying tropical timber will be discussed in more detail in Chapter 10.

Criteria for Assessing Trade Policy Options

The choice of which trade policy instrument is the most appropriate is broadly governed by:

- *conformity criteria* – ie whether the aims and means of each policy option are consistent with each other and with existing trade policies and trends; and
- *optimality criteria* – ie is each policy optimal in terms of achieving stated economic, environmental and social objectives.

The basic optimality criteria for assessing the trade policy options mentioned above may include:

- *Environmental effectiveness* This refers to the ability of the policy instrument to reduce timber-related tropical deforestation or at the very least to provide incentives for producer countries to ensure that their timber products come from 'sustainably' managed sources of supply.
- *Economic efficiency* This depends upon the relative economic benefits and costs of alternative policy instruments. This would take into account the economic implications of any creation or destruction of timber production and trade, and the level of financial resources devoted to implementing and enforcing the policy.
- *Distributional implications* This looks at the welfare effects of who gains and who loses from the policy intervention. Policy interventions may lead to a redistribution of finance between producing countries, consuming countries and the rest of the world, and along the production chain (eg forest owners, loggers, processors, traders, manufacturers, consumers). Some policy instruments have an explicit role in raising revenue to finance improved forest management which may have deliberate distributional implications.
- *Wider economic effects* This reflects the implications of the policy option for the economy, including balance of payment effects, and social impacts, such as employment implications and any knock-on effects (eg the implications for other bilateral trade flows).
- *Feasibility* The political acceptability of the policy both within a country and in the international community is crucial to whether it is adopted. In addition, the legality of the instrument (eg compatibility with GATT) will determine the feasibility of the policy option. The administrative and institutional capacity to implement the policy (which may include identification and certification, monitoring and enforcement) are also important.
- *Necessary and sufficient conditions* That is, to what extent will the policy instrument alone be sufficient in achieving the objective of sustainable forest management and, if not, what are the necessary conditions that are required to support it, eg labelling, enforcement, side payments and sanctions?

In an ideal world, all the above criteria would be weighed in order to determine which policy option is optimal. It is also possible that none of the trade policy options is satisfactory, in terms of meeting these criteria. In this respect we are unable to provide a comprehensive analysis of all the policy options as measured against the criteria. However, they are used in Chapter 10 as the basis for assessing the individual policy options as a guide to international policy seeking to develop a 'sustainable' management strategy – such as ITTO's Target 2000 (see Chapter 1).

Many of the timber trade policy options discussed above, such as bans, trade taxes or subsidies, quantitative restrictions, revenue-raising trade surcharges, and product labelling, essentially involve interventions in the tropical timber trade. A crucial issue in assessing these interventions is whether they ought to be pursued on a *unilateral* or *multilateral* basis. In Chapter 10, we assess each of the proposed trade policy options in more detail, drawing on the analysis and evidence assembled in this study. Before doing so, we want to discuss the general feasibility of employing unilateral and multilateral timber trade interventions to improve sustainable forest management.

THE FEASIBILITY OF UNILATERAL INTERVENTIONS

The timber trade interventions described in the previous section could easily be implemented *unilaterally*, ie either by one trading nation or a trading bloc. Indeed, there are strong pressures within some tropical timber importing countries to push their governments towards various unilateral interventions. Local government bans on the use of tropical timber have been implemented in some countries (eg Germany and the Netherlands). Product labelling has been implemented in Austria and is under consideration in Germany. The Netherlands has adopted a policy of importing only sustainably managed tropical timber by 1995, and is considering product labelling and quantitative restrictions as a means of achieving this objective. The United Kingdom is being urged by environmental groups to do the same. In all major Western importing countries, consumer-led boycotts of tropical timber – although not strictly a trade policy intervention – could further encourage official action.

The main objection to such unilateral policies is that they run the risk of violating internationally agreed trade rules, as negotiated under GATT. However, Article XX of GATT does allow some scope for the use of trade interventions for environmental objectives provided that such interventions do not 'constitute a means of arbitrary or unjustified discrimination between countries where the same conditions prevail or a disguised restriction on international trade'. More specifically, three sub-articles can be invoked on behalf of this exception:

- *Sub-article XX(b)* allows for trade measures undertaken 'to protect human, animal or plant life or health'.
- *Sub-article XX(g)* allows for trade measures necessary to the 'conservation of exhaustible natural resources if such measures are made effective in conjunction with restrictions on domestic production or consumption'.
- *Sub-article XX(h)* allows for trade measures undertaken 'in pursuance of obligations under any intergovernmental commodity agreement'.

Unilateral measures could only be justified on the basis of Sub-articles XX(b) and (g). Sub-article XX(h) clearly applies to multilateral measures alone.

Obviously, interpretation of these sub-articles for each specific case is required. Essentially, where a country needs to treat tropical timber imports differently from domestic goods in support of a regulatory scheme, Sub-articles XX(b) and (g) may be invoked, provided the measure is not unjustifiably discriminatory nor intended to inhibit trade.

Sub-article XX(g) does allow some latitude for unilateral trade restrictions in support of 'conservation of exhaustible natural resources', which may be increasingly invoked for intervention in the timber trade. So far, rulings by GATT panels have involved fairly narrow interpretations of this sub-article. For example, attempts by the United States to impose a ban on tuna imports from Canada in 1979 and on imports of yellow fin tuna and tuna products from Mexico in 1990 were both invoked with reference to Sub-article XX(g). In its case against Canada, the US argued that: tuna was an exhaustible and over-exploited resource; the prohibition was not discriminatory in that similar measures had been taken for similar reasons against imports from other countries; the action was taken in conjunction with measures aimed at restricting domestic production or consumption although not specifically against albacore tuna; and the measure was related to conservation of tuna in that it was taken to prevent threats to the international management approach that the US viewed as essential to the conservation of tuna stocks.

However, the GATT panel ruled that the ban could not be justified under Sub-article XX(g) because the United States had imposed the ban against imports of all tuna and tuna products, whereas restrictions on domestic catch had so far been applied only to specific tuna species (Grimmett 1991). Similarly, the US embargo of all yellow fin tuna and tuna products harvested with purse seine nets in the eastern Pacific Ocean by tropical foreign nations was justified in terms of the need to protect Pacific dolphins. However, the GATT panel recently upheld Mexico's complaint that such an action was discriminatory.

There is considerable debate as to whether similar environmental controls, such as a ban on tropical timber imports, are compatible with GATT Sub-article XX(g). In reviewing the feasibility of the Netherlands adopting its 1995 policy, Hewett, Rietbergen and Baldock (1991) argued that a Dutch ban on imports of non-sustainably produced tropical timber could comply with the Sub-article if:

- restrictions on domestic consumption of tropical timber were introduced (eg in construction or furniture)
- the controls were applied strictly in line with the criteria of sustainable management of production, and avoided discrimination between countries on other grounds
- the reasons for the ban were made explicit, allowing sustainably produced timber unrestricted entry, and
- if the ban can be justified and explained in terms of conserving tropical forests.

However, others are more pessimistic, and expect any import ban or other environmental control on the timber trade not only to be unacceptable under recent GATT rulings but, in the case of the Netherlands or the United Kingdom adopting a unilateral policy, also to contravene EEC legislation (Arden-Clarke 1991; Dudley 1991).

Recent calls for broadening the objectives of the GATT – or of its possible 'successor' the World Trade Organization (WTO) – to further the use of trade interventions for environmental policies will, if successful, open the door to a whole host of unilateral timber trade measures. For example, Arden-Clarke (1991) has argued that the GATT should be amended to allow any of the 'external' environmental costs of production to be considered essentially an environmental 'subsidy' for any output that is subsequently exported. Thus GATT should be amended to allow contracting states to:

- discriminate between like products that vary in the degree to which the environmental and resource costs of their production are incorporated in their price
- protect domestic industries that internalize more of their costs than foreign competitors, with import tariffs or export subsidies, and
- provide subsidies to maintain the competitiveness in international markets of exported products with greater cost internalization than competing products.

Such an approach would clearly allow nations to invoke the full range of policy measures available to them for intervention in the timber trade. For example, the United States has already imposed environmental restrictions on its own Pacific North-west logging operations.[1] Given this policy, and under amended GATT rules, the US Government would be able to limit timber imports from other countries that did not maintain similar environmental standards for their timber operations. Alternatively, the US could choose to subsidize plywood and lumber mills in the region and elsewhere in the country for their loss of 'international competitiveness' as a result of the environmental restrictions.

In addition, under the above amended GATT rules, any importing country or trading bloc could impose trade sanctions on producer countries that appeared to be producing timber or timber products 'unsustainably', particularly as the result of the type of policy failures described in Chapter 6. To the extent that 'unsustainable' management leads to lower harvesting costs and thus lower export product prices, importing countries would have the right to impose tariffs as a means of correcting this 'environmental' export subsidy.

Whether a broadening of the GATT rules is achieved or not, or whether further 'greening' of trade occurs through the new WTO, some countries and trading blocs are considering adopting stringent interventions in the international timber trade on environmental grounds – such as a ban on imports of tropical timber that is not 'sustainably' produced. For example, as mentioned above, the Netherlands has recently declared a policy whereby from 1995 (or earlier if the 1993 evaluation shows this to be feasible) all tropical hardwood imports must originate from countries where logging is subject to national/regional long-term forest management plans, and where adequate provision has been made for the protection of largely

---

[1] For further discussions of this policy, particularly its implications for the world timber trade, see Barbier (1994); Flora and McGinnis (1991) and Perez-Garcia (1991).

virgin tropical forests, and for the conservation and sustainable management of other forests (Government of the Netherlands 1991).

In bilateral consultations and international fora (eg the EEC, ITTO and the FAO Tropical Forest Action Plan), the Netherlands will also seek agreement to adopt the same approach, and assistance in achieving this objective. For example, through:

- an international labelling system to distinguish 'sustainably' produced timber commodities; protection for species threatened with imminent extinction;
- financial support to tropical forest countries for the development and implementation of long-term timber production and management plans, and their monitoring and evaluation; and,
- generally accepted guidelines on the granting of and conditions governing timber concessions, and on appropriate monitoring.
- The Dutch advocate raising the necessary financing for this policy internationally, through an import levy on tropical timber, supplemented by compulsory government contributions in proportion to commercial interest.

The main thrust of the Netherlands' policy is clearly to stimulate complementary actions at the international level, eg through the EEC and ITTO. For example, in considering a similar policy for the UK, Dudley (1991) has suggested that its most important impact would be to send 'a clear message both to exporters and to other importing countries regarding the urgency of developing sustainable forestry in the tropics'. Nevertheless, there are some important obstacles to such unilateral actions.

First, such import interventions could be interpreted as *discriminatory*, in the sense that only tropical, and not temperate, timber producing countries would be affected. There is little evidence to suggest that temperate production of timber is necessarily 'sustainable', yet any tropical timber import interventions would exclude temperate wood products.[2] This seems particularly misplaced, given that the largest hardwood reserves are in the temperate regions of the United States, Canada and Russia (see Chapter 2). On the other hand extension of import interventions to all timber products would most likely invite retaliatory action from those developed market economies that produce temperate timber. Developing countries would argue, with some justification, that their ability to retaliate in response to a tropical timber import intervention is much more limited, which only highlights the discriminatory nature of such unilateral action.[3]

Second, imposition of tropical timber import interventions on the grounds of 'sustainable management' may prove to be *arbitrary* and possibly *unworkable*.

---

2 For reviews of temperate forest policy and trade with regard to sustainable management, see Barbier (1994) and Wibe and Jones (1992).

3 The Dutch policy in so far as it restricts imports of tropical timber may in particular be accused of protectionism, as at the same time the Netherlands also 'seeks to increase domestic timber production . . . The policy may have an indirect impact on the rainforests, since the more self-sufficient the country becomes, the less tropical timber it will need to import. The plan is to increase forest acreage in the Netherlands by 38,000 ha from 1977 to 2000' (Government of the Netherlands 1991, p 22).

As argued by Poore et al (1989), 'it is questionable whether one universal definition of "sustainable management" is useful, because it will lend itself to different interpretations by different interests'. Yet, unless there is an international consensus on what is meant by 'sustainable management', it will be left to each nation or trading bloc to determine its own criteria, which may differ markedly. Attempts by ITTO to develop guidelines on sustainable management suggest a total of 41 principles and 36 recommended actions covering forest policy, management, and socio-economic and financial aspects (ITTO 1990). Much more work is required on such guidelines by ITTO and other institutions before they are universally acceptable for implementation.

Similar problems exist for labelling 'sustainably' versus 'unsustainably' produced products. Certification would have to be carried out by an independent body with sufficient standing to win the trust of importers and consumers, and with a reputation of its own which is important enough to guard against corruption. A network of qualified inspectors would be needed to make site inspections and keep careful accounts of the amount of timber coming out of certified forests and plantations, and there would have to be penalties for fraud and negligence (Dudley 1991). On the other hand, producer countries would object strongly to such interference with domestic forestry policy and operations, and would be less willing to cooperate in such an operation if import restrictions and certification are imposed unilaterally.[4]

Third, a unilateral trade intervention may also be *ineffective* in reducing either tropical deforestation or the trade in 'unsustainable' timber. As discussed in Chapters 2 and 3, timber production is not the major cause of tropical deforestation, not all (and a declining share) of the tropical timber produced is for export, and an increasing share of tropical timber exports is being absorbed in South–South trade. This would suggest that, in response to unilateral interventions imposed by importers, major tropical timber exporters (eg in SE Asia) may be able to divert some timber supplies to domestic consumption, or to other export markets, fairly easily. For those tropical forest countries where timber exports are not significant, and are not a major factor in deforestation (eg Latin America), a unilateral ban – even if imposed by a major trading bloc – would have little impact on timber management or overall deforestation.

In fact, any major tropical timber import intervention may encourage acceleration of the kind of distortionary export policies discussed in Chapter 7. Tropical forest countries could justify this policy by the need to 'compensate' domestic processors for the distortionary impacts of the import policy and to 'squeeze' more value added out of their remaining exports. In addition, processed timber products would be more difficult to 'certify' as to whether they have been produced 'sustainably' or unsustainably', facilitating the evasion of the import intervention.

Finally, a unilateral intervention might have little impact on the economic *incentives* for sustainable management at the concession level, and may actually encourage poor management practices. As discussed in Chapter 6, a host of domestic and market policy failures in tropical forest countries affect the

---

4 For further discussion of the issues concerning certification see Chapter 10.

'internalization' of the user and environmental costs of timber harvesting by concessionaires. Major policy changes in the forestry sector will be required to address these issues, yet by imposing a unilateral trade intervention, an importing country may reduce any political leverage it could have to influence policy makers in producing nations. Thus Hewett, Rietbergen and Baldock (1991) argue that any action such as a unilateral tropical timber import ban must be accompanied by 'positive incentives' to producer countries, eg financial aid to facilitate the changes needed to progress towards more sustainable timber management and compensation for short-term losses in export earnings. However, such bilateral transactions do not guarantee that improvements in incentives at the concession level will be forthcoming, unless monitoring, certification and penalties can be strictly and efficiently enforced.

Moreover, as an extension of the 'rent capture' argument outlined in Chapter 6, a bilateral transfer of funds at the *government* level may do little to alter incentives perceived by the *concessionaire* to improve timber management practices. On the contrary, as the concessionaire has to bear the costs of altering behaviour but receives no financial gain, the incentives for high-grading, trespassing and avoiding taxes and regulations are increased, especially for infra-marginal stems and stands.[5]

So far, we have discussed unilateral action from the standpoint of import interventions in the tropical timber trade. It is also conceivable that some 'unilateral' policy is also adopted at the producer end of the trade. The possibility of a large country exporter (or a group of exporters) of tropical timber products using its (their) 'market power' to influence international prices is explored by Barbier and Rauscher (1994) in a theoretical model. Their results indicate that if a large country exporter can increase its market power, and influence international timber product prices, then it can afford to conserve more of its tropical forests in the long run. Increased monopolistic power or cartelization of tropical timber supply appears to have a resource-conserving effect. If the large country or cartel also receives international financial assistance for sustainable forest management and conservation, then even more of the resource will be conserved.[6]

However, in practice there are obvious problems with tropical timber producing countries forming a cartel and asserting market power. As discussed in Chapter 2,

---

5 Essentially, in exchange for the financial aid, the tropical forest government would impose additional 'sustainable management' regulations or taxes, which would increase the marginal costs, and decrease rents, to the concessionaire. However, on inframarginal stands and stems, ie where economic rents or returns are greater than zero, the concessionaire would have an incentive to highgrade, trespass and otherwise avoid these regulations, and probably could afford to do so because of the excess profits.

6 It should be noted, however, that this theoretical result derived by Barbier and Rauscher (1994) is influenced by whether or not the tropical forest country takes into account fully the social (non-timber) benefits of the forest stock and that harvesting is both privately and socially efficient. As indicated in Chapter 6, it is clear that the policies and management strategies of most major tropical forest countries do not conform to this ideal.

current trends suggest that tropical timber does not dominate the international timber market, and those tropical timber exporters (eg Malaysia and Indonesia) that are significant in the world timber trade may not be able to maintain their supremacy over the long term.[7] In addition, cartelization is always a major problem politically and practically, especially in the tropical timber trade where so many different types of wood and products are traded.

Finally, there is no guarantee that cartelization by itself would affect the *incentives* for sustainable management at the concession level. As argued above, an increase in funds and revenues at the *government* level may do little to alter the incentives of a *concessionaire* to improve timber management practices, and may actually encourage poor management practices. In fact, the theoretical results derived by Barbier and Rauscher (1994), which demonstrate that monopoly power will lead to forest conservation over the long run, depend crucially on the assumption that the policy makers are interested in conserving the resource and that policies for sustainable timber management are already in place.

Other unilateral export interventions to improve sustainable management, such as export tax surcharges, taxes on 'unsustainably' produced exports, and export subsidies for 'sustainably' produced exports, also presume that: i) 'political will' to support sustainable timber management already exists; and, ii) the resulting sustainable management practices and regulations are implementable and enforceable. Unlike unilateral import interventions, there would be fewer international disputes over such unilateral export policies, as they would presumably not lead to an obvious unfair trade advantage by the tropical timber exporting country adopting the interventions. If anything, export market shares and earnings may be negatively affected in the short run. The exception might be a subsidy for 'sustainably' produced timber exports, as the policy could easily be a disguised means of trade promotion. International competitors could insist that a tropical timber exporting country adopting this policy unilaterally demonstrate unequivocally that it is subsidizing sustainably managed timber only and that the additional costs of producing this timber justifies such a subsidy.

In sum, the pressures for importing countries in particular to adopt unilateral interventions in the tropical timber trade are mounting. Several governments appear already to be responding to this pressure, and more are likely do so in the near future. However, as this section has argued, such interventions tend to be discriminatory, arbitrary and possibly unworkable. They may also prove to be ineffective in reducing either tropical deforestation or the trade in 'unsustainable' timber, and may do nothing to encourage economic incentives at the concession level for sustainable management. The conclusion must be that unilateral timber trade interventions are not the appropriate means of ensuring sustainable management of tropical timber production.

---

[7] For example, see the projections for this study of the CGTM in Perez-Garcia and Lippke (1993)

THE FEASIBILITY OF MULTILATERAL INTERVENTIONS

If unilateral timber trade interventions are not appropriate or acceptable internationally, then some view multilateral interventions as the only solution. In fact, as discussed above, some of the unilateral actions are essentially geared towards stimulating complementary or concerted actions at the international level. However, it is worth noting that the main impetus for multilateral action is the perception that tropical forest management is a *global* issue that involves *global* agreement and action.

Mounting scientific evidence suggests that forests may play an important role in global climate regulation. Accelerated deforestation over the past century has been implicated in global warming, changing weather patterns and the rise in mean sea levels. It has also become clear that tropical forests, in particular, harbour a disproportionate share of the world's natural wealth of genetic resources (biodiversity). These resources are of uncertain but possibly significant economic value, not only in terms of their potential uses in agriculture, industry and medicine, but also simply in terms of the value placed on their continued existence by the general public (existence value). Increasing publicity and education about the intricate and fragile ecology of tropical forest areas has led to growing concern for their preservation.

These phenomena imply that tropical forests have an important global economic value, which is not captured by producers of tropical timber products and is therefore not reflected in the market prices of these products or their derivatives. Any significant timber-related tropical deforestation inevitably results in some destruction of this global value. This *global externality* has been identified as a possible justification for new and/or increased international financial transfers, in favour of nations harbouring large tracts of tropical forest land. The idea is that all nations would benefit from increased efforts to preserve the world's remaining major tropical forest areas. All nations, it is argued, should therefore contribute towards *compensating* producing nations for the loss of potential income that they would incur in reducing tropical deforestation, timber sales and conversion of forest land to other uses.

As discussed in Chapter 5, these concerns, as well as the desire to see traded timber produced more sustainably, are the main motivations behind the impetus for increased multilateral interventions in the timber trade. Moreover, if multilateral interventions in the tropical timber trade are to be implemented, then the International Tropical Timber Agreement (ITTA), and its executive body the ITTO, would be the most appropriate policy forum for seeking agreement from representatives of major consumer countries, producer countries, timber traders and NGOs.

At the 10th Session in May 1991, the ITTC unanimously adopted a 'Year 2000 Target' that encourages 'ITTO members to progress towards achieving sustainable management of tropical forests and trade in tropical forest timber from sustainably managed sources by the year 2000' (Decision 3(X) ITTC 1991). By adopting this target, ITTO is clearly within GATT rules, as indicated by Sub-article XX(h). The target has major political significance as, under the auspices of the ITTA, producer countries, consumer countries and international timber traders have to cooperate in taking practical steps towards sustainable timber management and production.

However, by no means does the Year 2000 Target imply that the ITTA would necessarily endorse the use of even multilateral interventions in the tropical timber trade. On the other hand, if all parties to the ITTA did feel that such interventions were warranted, then most likely they would be accepted under even existing GATT rules. [8] Sub-article XX(h) is fairly clear that it would allow for trade measures undertaken 'in pursuance of obligations under any intergovernmental commodity agreement'.

Thus, assuming that they are adopted under ITTA auspices, any of the trade policy options reviewed above that essentially call for multilateral trade interventions could be considered *feasible* – in the narrow sense of not being *discriminatory* under existing international trade rules. Whether some of the trade interventions are *desirable* or *appropriate* from the standpoint of encouraging sustainable timber management is another matter. For many of the interventions, problems of *workability*, *effectiveness* and *incentives* still remain.

CONCLUSIONS

This chapter has identified a number of possible tropical timber trade policy options that could be implemented to influence the incentives for sustainable production forest management. Many of these policy options involve significant interventions in the tropical timber trade, which would have varying degrees of distortionary effects. Although Chapter 10 will assess all the trade policy options in more detail, this chapter has discussed the general feasibility of implementing timber trade interventions either unilaterally or multilaterally. We conclude that:

- Although the pressures for importing countries in particular to adopt unilateral interventions in the tropical timber trade are mounting, such interventions are not the appropriate means of ensuring sustainable management of tropical timber production.
- Unilateral interventions tend to be discriminatory, arbitrary and possibly unworkable. They may also prove to be ineffective in reducing either tropical deforestation or the trade in 'unsustainable' timber – and may in fact be counter-productive. In addition, they do nothing to encourage economic incentives at the concession level for sustainable management.
- Assuming that they are adopted under ITTA auspices, multilateral trade interventions could be considered feasible – in the narrow sense of not being discriminatory under existing international trade rules. Whether some of the trade interventions are desirable or appropriate from the standpoint of encour-

---

8   In addition, similar national initiatives, such as the trade interventions implied in the Dutch policy of requiring all tropical timber imports to be 'sustainably' produced by 1995, may no longer be dismissed as unilateral initiatives, although action could still be taken against the Netherlands for 'jumping the gun' by five years.

aging sustainable timber management is another matter. For many of the interventions, problems of workability, effectiveness and incentives still remain.

In the next chapter, we assess each of the trade policy options in more detail in order to determine their appropriateness for moving the tropical timber trade towards improved sustainable management of production forests. By making this assessment we hope to clarify which multilateral interventions in the tropical timber trade – if any – can provide workable and effective incentives for sustainable management.

10
ASSESSMENT OF TRADE POLICY OPTIONS

Chapter 9 identified several tropical timber trade policy options to promote sustainable forest management and the basic criteria for assessing them. In this chapter, we draw upon the various analyses conducted for this book in order to assess each trade policy option. The policy options are grouped into three main categories:

- do nothing;
- take measures to alter the pattern of tropical timber trade; and,
- take measures to raise revenues for sustainable forest management.

DO NOTHING

The option of 'doing nothing', ie allowing existing tropical timber trade policies to remain as they are today *and* not implementing addition policy interventions to affect trade patterns and/or the incentives for sustainable forest management, should always be considered a serious policy choice. For example, 'doing nothing' would be the appropriate policy choice if tropical deforestation was not perceived to be an economic problem. 'Doing nothing' would also be attractive if other policy options – ie 'doing something' – appear ineffective or undesirable, or if the costs of these options outweigh the benefits.

Another obvious attraction of the 'do nothing' option is that it does not present any administrative and institutional obstacles for consumer and producer countries, which would clearly not be the case for any new trade policy initiative. Similarly, no new support mechanisms, such as certification, enforcement, side payments and sanctions, would have to be devised.

However, 'doing nothing' is not considered an attractive policy option given current trends in timber trade and tropical deforestation, and above all, given that sustainable management of production forests is still not widespread in tropical forest countries. Chapters 2 and 3 have pointed out that, although timber production is not the main source of tropical deforestation, logging efforts have clearly made some contribution to the problem. More importantly, by adding value to forestry operations, the trade *could* act as an incentive for sustainable production forest management – provided that the appropriate domestic forestry policies and regulations are implemented. As argued in Chapter 3, unless it can earn sufficient economic returns, sustainable timber management will be unable to compete with alternative uses of forest land, such as conversion to agriculture. This may lead to higher rates of tropical deforestation in the long run.

Chapters 6–8 have discussed a number of existing domestic and trade policy distortions that affect both the trade and timber-related deforestation. In particular, domestic economic and forestry policies in tropical timber producing countries have tended to distort the incentives for sustainable management of timber resources and have failed to curb the wider environmental problems associated with timber extraction. Trade policy distortions in producer and consumer countries have, if anything, exacerbated this situation. Again, this would suggest that current policies affecting tropical timber production and trade are not providing the appropriate incentives for sustainable production forest management. Leaving these policies as they are, therefore, is not an appropriate solution.

Moreover, current and future market trends for tropical timber products are not likely to provide incentives automatically for sustainably managed tropical timber. As discussed in Chapters 2 and 4, producers of tropical timber products must increasingly compete with each other and with producers of temperate wood products for shares of consumer markets. Although tropical timber resources are being depleted, their impact on global timber production and trade will be offset by expanding temperate resources, non-timber substitutes, and technical improvements in efficient resource production and processing. As a result, real tropical timber product prices in major consumer importing markets are not projected to increase significantly in the short term (see Chapter 2 and Perez-Garcia and Lippke 1993). Inventory constraints in producing countries may push up log prices in these countries in the medium and long term. However, in the *absence of domestic policies within tropical forest countries to increase the incentives for sustainable forest management*, any subsequent increase in prices at the timber stand level will not reinforce these incentives. Thus timber-related deforestation may continue to be a problem.

In sum, there is very little reason to believe that 'doing nothing' is an appropriate response to the current situation. In particular, this option is unlikely to lead to greater trade-related incentives for sustainable management of tropical production forests.

MEASURES TO ALTER THE PATTERN OF TRADE

In Chapter 9, several policy interventions that would alter the pattern of tropical timber trade significantly were identified. These include:

- trade liberalization;
- trade bans;
- quantitative restrictions;
- trade taxes; and,
- trade subsidies.

Each of these policy options is assessed briefly.

Trade Liberalization

Removal of existing interventions in the trade at the export and import end were reviewed separately, in Chapters 7 and 8 respectively. In addition, the CGTM for forest products was used to simulate the effects of complete removal of import product tariffs and log export bans (see Perez-Garcia and Lippke 1993).

Chapter 7 noted that the export restrictions recently imposed by timber producing countries have not really been adopted to reduce timber-related deforestation. The reductions in deforestation have not been dramatic in the short term. The expansion in processing capacity and the problems with inefficiencies and overcapacity may have exacerbated timber-related deforestation over the medium and long term. However, this does not imply that the removal of these export restrictions will automatically lead to an increase in the incentives for sustainable forest management. The available evidence suggests that trade liberalization could in fact *increase* the pressure on the resource base, *unless well-enforced sustainable forest management policies and regulations are also implemented*.

Chapter 8 indicated that existing import barriers to tropical timber products in developed market economies do depress demand for these products somewhat. The main problem appears to be *tariff escalation* – the tendency for import tariffs to rise with the degree of processing of the product.

Tariff reduction by developed market economies would most likely lead to trade *creation* rather than trade *diversion*, with almost all new trade created for developing countries involving markets in the EEC, Japan and the United States. However, it is also not clear that timber-producing countries would be the main beneficiaries from the additional trade created. As discussed in Section 8, NICs that have no or negligible production forests, such as Taiwan, South Korea and Singapore, currently have an advantage in more advanced processed products, such as plywood and veneer. Thus current tariff barriers may be discriminating more against the NICs than the timber-producing countries.

Chapter 8 also pointed out that non-tariff barriers facing tropical timber products in developed market economies are much more significant than tariff barriers. However, it is very difficult to determine the effects of non-tariff barriers on the trade, and consequently, what impact their removal might have. Similarly, although the GSP may reduce tariffs for developing countries, the tariff quota/ceiling system operated by the EEC and Japan effectively imposes quantitative restrictions on key imported tropical timber products, notably plywood. Removal of this specific quota system, as proposed in the recent GATT negotiations, would probably have more of an impact on the trade in tropical timber products than a general 'trade liberalization' through tariff reduction.

In addition, the tariff and other import barriers in developing countries for timber products are generally much higher than those in developed market economies. For example, some major exporting regions (eg Indonesia and peninsular Malaysia) completely ban the import of timber products. Complete liberalization of the trade would mean the removal of these import barriers as well, and by no means would it be certain that there would be a net gain for these exporters.[1]

The CGTM trade liberalization policy simulation examined the implications of a 10 per cent across the board reduction in 'transfer costs', ie the difference between export prices and import prices, as a proxy for the removal of both tariff and non-tariff barriers to tropical timber products. The simulation also included the complete removal of log export bans from Malaysia west (includes peninsular Malaysia), Indonesia, the Philippines, Papua New Guinea and West Africa.[2] The main results indicate that:

- The reduction of tariffs and the elimination of bans increases global demand for tropical hardwoods, with significant producer log price increases of about 20 per cent.
- Consumer countries also benefit, as the increased availability of tropical timber logs internationally causes a significant decline in the prices of imported logs.
- However, as the tropical timber producing countries have little comparative advantage in processing, they also lose some of their processing capacity to their major importers.
- As a result, there is a major shift in the pattern and distribution of tropical timber product exports from the major Asian exporting regions, ie more log and less plywood exports. The exception is Malaysia west, which becomes a substantial log importer in order to expand plywood exports.[3]

Thus the policy scenario suggests that trade liberalization would most likely produce significant gains for importing countries, particularly those with log processing capacity. The impacts on tropical timber exporting countries are more mixed. In the policy scenario, the rise in producer log prices could act as an important incentive for sustainable timber management – but the scenario assumes that by the year 2005 *policies promoting sustainable harvest levels are in place* in Indonesia and Malaysia. Without such policies, higher prices could conceivably lead to increased 'mining' of remaining commercial timber reserves in those countries.

However, the main obstacle in recommending general trade liberalization for tropical timber products as a policy option is that it may simply not be politically realistic in the current global trade climate. Getting exporting and importing countries to agree to such wide-sweeping reductions in existing tropical timber trade restrictions would be a very tall order. For example, as noted in Chapter 8, GATT negotiations for removing just one import restriction – the EEC plywood quota system – failed to obtain agreement on specific actions. Exporting countries

---

1. As discussed in Chapter 7, part of the problem is that Malaysia and Indonesia have been using their export and import restrictions as a means to increase their market power, notably to capture larger shares of the international non-coniferous plywood market. Most likely, complete trade liberalization would undermine this monopoly position.

2. See Perez-Garcia and Lippke (1993) or Annex K in Barbier et al (1993b) for further details.

3. This reflects the imminent depletion of commercial inventory in the region, forcing Malaysia west to import logs to supply its processing industry; see Perez-Garcia and Lippke (1993).

would also be extremely reluctant to remove export restrictions on logs, particularly if it would mean loss of processing capacity to importing countries.

In sum, complete removal of all export and import restrictions on the tropical timber trade is simply not feasible and, in any case, may not be environmentally desirable unless accompanied by improved forestry policies and regulations. For example, the extent of non-tariff barriers and their impacts on the trade are simply not well understood. However, even eliminating the more 'visible' quantitative restrictions and tariffs across all countries is probably not realistic either. Both importing and exporting countries will most likely continue to employ trade restrictions as part of their national strategies for forest-based industrialization and protection of domestic industries. Thus the 'political will' for a general liberalization of the tropical timber trade may simply not be there.

Nevertheless, there are some selective 'trade liberalization' steps that could be promoted, if supported through a multilateral agreement such as the ITTA and implemented collectively by ITTO on behalf of its member countries:

- First, ITTO could request that importing countries revise policies that clearly discriminate against tropical timber exporters, such as the TQ/ceiling system of EEC and Japan for non-coniferous plywood exports. For example, ITTO could suggest that exporting countries that *show demonstrable progress towards achieving Target 2000 of sustainable timber management* could be allowed exemption from the quota penalties of such systems. The same principle could be applied to other tariff and non-tariff barriers operated by developed market economies on a case by case basis.

- Second, ITTO could request that exporting countries conduct thorough national reviews of the impacts of their trade restriction policies on sustainable timber management. In particular, the reviews should focus on the extent to which the trade restrictions exacerbate problems caused by poor domestic forestry regulations and policies. ITTO could recommend that these trade restrictions only be continued if: i) they do not appear to be contributing to greater timber-related deforestation; and, ii) if the exporting country *demonstrates progress towards achieving Target 2000 of sustainable timber management*, most notably by implementing well-enforced sustainable forest management policies and regulations.

Trade Bans

As discussed in Chapter 9, pressures from environmental groups and consumer-led boycotts in developed market economies is leading to serious consideration of a *complete* ban on the import of all tropical timber products, or at least a *selective* ban on those products that are not 'sustainably produced'.

The results of our study suggest that, despite their popular appeal, the use of such bans would not be appropriate for encouraging sustainable management in tropical timber exporting countries. There are several reasons for this.

First, producer countries argue, with some justification, that a ban on tropical timber products is *discriminatory*; ie similar rules do not apply to the temperate timber trade. It is unlikely that they would allow such a policy option to be sanctioned by any multilateral forum such as the ITTA and would certainly oppose it as contravening GATT. As will be discussed below, without the cooperation of producer countries, an import ban would most likely be *arbitrary* and *unworkable*.

To extend the ban to include both tropical and temperate product trade would be even more unfeasible. Those developed market economies that produce temperate timber also tend to be the most under pressure from environmental groups to instigate a tropical timber import ban. Given that the global temperate market is much larger and highly competitive, the governments in these countries would probably resist hurting the prospects of their own forest industries by extending the ban to cover all timber product trade.

More importantly, there is substantial evidence that an import ban on tropical timber products would be *ineffective* in reducing either tropical deforestation or the trade in 'unsustainable' timber – and may be even *counter-productive*. As discussed in Chapters 2 and 3, timber production is not the major cause of tropical deforestation. Not all (and a declining share) of the tropical timber produced is for export and an increasing share of tropical timber exports is being absorbed in South–South trade. This would suggest that, in response to a tropical timber ban imposed by current importers, major tropical timber exporters (eg in SE Asia) may be able to divert some timber supplies to domestic consumption, or to newly emerging export markets, fairly easily. For those tropical forest countries where timber exports are not significant, and are not a major factor in deforestation (eg Latin America), a ban may have little impact on timber management or overall deforestation.

In fact, to the extent that a tropical timber import ban does affect the export of tropical timber products substantially, it would have little impact on the economic *incentives* for sustainable management at the concession level, and may actually encourage poor management practices. As discussed in Chapter 6, domestic and market policy failures in tropical forest countries affect the 'internalization' of the user and environmental costs of timber harvesting by concessionaires. Major policy changes in the forestry sector will be required to address these issues, yet by imposing trade bans importing countries may reduce their *political leverage* in influencing policy makers in producing nations.

Although in the short run a trade ban may reduce pressure on tropical forests through lower production for the trade, in the medium and long run a ban is likely to have a detrimental impact. By eliminating the gains from trade, a ban on tropical timber imports would decrease the value derived from timber production and thus actually reduce the incentives for tropical forest countries to maintain permanent production forests. Faced with declining export prospects and earnings from tropical timber products, developing countries may decide to convert more forests to alternative uses, notably agriculture. Thus the effect of the ban may be to reduce log production and exports, but may actually increase overall tropical deforestation in the medium and long term.

As part of the analysis conducted for this study, we examined the effects of a total import ban on a major tropical timber producer, Indonesia (see Barbier et al 1993a). According to the results of the analysis, an import ban would have a devastating impact on Indonesia's forest industry in the short term. Although there would be significant diversion of plywood and sawnwood exports to domestic consumption, this would be insufficient to compensate for the loss of exports. Net production in both processing industries would fall. Given its export orientation, the plywood industry would be particularly hurt – reducing its output by over 40 per cent. Net production losses in the sawnwood industry would be closer to 10 per cent. The overall effect is to lower domestic log demand in the short term by around 25–30 per cent.

The policy scenario is of course assuming that the import ban is 100 per cent effective. It is unlikely that all importers of Indonesia's tropical timber products – many of which are also newly industrializing or producer countries with processing capacities – would go along with a Western-imposed ban. In any case, one would expect that over the longer term there would be some diversion of Indonesian plywood and sawnwood exports to either new import markets or existing markets that prove to be less stringent in applying the ban (eg other developing or newly industrializing countries).[4]

In addition, the long-term implications of an import ban on tropical deforestation are also uncertain. Even if the ban is 100 per cent effective in the short term, any reduction in tropical deforestation resulting from lower levels of timber harvesting is likely to be shortlived. There are several reasons for this.

First, diversion of timber products to satisfy domestic demand is likely to continue as Indonesian population and economic activity expand (Barbier et al 1993a). Thus one can expect the wood processing industries to recover somewhat through reorientation to meet local consumption of timber products. Moreover, past evidence suggests that domestic-oriented processing in Indonesia is less efficient than export-oriented processing, implying poorer wood recovery rates and greater log demand (Constantino 1990). Pressure on the tropical hardwood forests may therefore increase after the initial 'shock' of a ban.

However, it is unlikely that the domestic market in Indonesia will generate the same demand for higher valued timber products as the international market. Instead, domestic demand is likely to be strongest for high-volume but lower valued wood products. As a consequence, it may be difficult for Indonesia to justify holding as large a proportion of its tropical forests as permanent production forests if the expected economic returns from sustainable management decline as a result of the ban. More of the resource may be shifted to conversion forests, and timber production will become residual to satisfying the growing demand for agricultural land. In short, without the timber trade providing increased value added in the form of external demand for higher valued products, there may be less reason for Indonesia 'holding on' to these forests as opposed to converting them to an alternative use, such as agriculture (Barbier 1993; Vincent 1990).

To summarize, a total import ban would cause a major diversion of Indonesian timber products to meet domestic demand. Although in the short term net production of wood products, and thus log demand, would fall, this situation would not necessarily be sustained over the long run. Even if this is not the case, the ban may be ineffective in permanently reducing tropical deforestation because:

■ timber production is not the main source of deforestation in Indonesia; and,

---

4   As discussed in Barbier et al (1994), these effects cannot be captured in our model.

- as the value of holding on to the forest for timber production decreases, the incentives to convert more of the resource to alternative uses such as agriculture will increase.

Many of the problems associated with a *complete* import ban on tropical timber products would also apply to a *selective* import ban on 'unsustainably' produced timber alone. In our opinion, a selective import ban would also fail to provide the appropriate economic incentives for sustainable management at the forest level, and may also be counter-productive, for the similar reasons outlined above for a complete ban, namely:

- *Diversion of trade to other markets* (domestic, export markets without bans etc). If these markets are for lower value products, producer countries may need to supply higher volumes of tropical timber to generate substantial earnings, thus leading to more pressure on timber resources.
- *Lower political leverage of importing countries* to influence forestry policies in producer countries.
- *Little positive reinforcement of the incentives for sustainable management.* Selective bans would have an immediate impact on a country's ability to derive value from timber production, and would act as a disincentive in the medium and long term to maintaining tropical forests for timber production, as opposed to conversion to agriculture and other uses.
- *Generate incentives to circumvent the ban.* There is generally high elasticity of substitution for tropical timber products from different sources of origin, particularly for higher valued products such as plywood (see Chapter 4). Thus producer countries would gain significantly if they could 'pass off' their 'unsustainably' produced timber as 'sustainably' produced.

Moreover, as discussed in Chapter 9, selective bans have the additional complication of the need for a *non-arbitrary* and *workable* international certification process to distinguish 'sustainably' versus 'unsustainably' produced tropical timber products.[5] Given the lack of data on forest inventories and offtake levels, such certification would be extremely difficult to establish in the near future without the cooperation of producer countries. However, there is little incentive for producer countries to participate in this process. Without this data, a selective ban could not possibly succeed. In addition, an effective monitoring and evaluation system would be required to enforce a selective import ban. No such mechanism currently exists or is likely to be implemented in the next few years. Again, cooperation by producer countries, which would be fundamental to the success of the verification system, is unlikely to be forthcoming. Finally, a selective trade ban is unlikely to be acceptable in the international political arena – particularly as producer countries will dismiss it as *discriminatory*.

On existing evidence, we conclude that neither a complete nor selective import ban on tropical timber products will provide appropriate incentives for sustainable management of production forests, and may even be counter-productive by en-

---

5 As a number of trade policy options require a similar process in order to be effectively implemented, we discuss the issue of *certification* separately below.

couraging the conversion of these forests and any remaining 'virgin' forest areas to alternative uses.

Quantitative Restrictions

As discussed in Chapter 9, to some extent quantitative restrictions are similar to trade bans, but usually take a less severe form. However, a 100 per cent restriction on the quantity of trade is effectively the same as a complete ban on all international trade in tropical timber. Thus, to a large extent, many of the problems associated with tropical timber trade bans also apply to quantitative restrictions.

Two quantitative restrictions are worth briefly examining:

- quotas targeted to specific product categories, such as timber products derived from endangered species; and,
- quotas conditional on the level of sustainable offtake of specific species.

At the 8th Conference of Parties to the Convention on International Trade in Endangered Species (CITES), several tropical timber species were proposed for the Appendix I and II list of CITES (ITTC 1992b). Species given an Appendix I listing are effectively banned from commercial trade, or only authorized 'in exceptional circumstances'. Species listed on Appendix II are subject to strict regulation 'in order to avoid utilization incompatible with their survival'. Of the species proposed for listing at the meeting, *Intsia spp* (merabu), *Gonystylus bancanus* (ramin), *Swietenia spp* (mahogany) and *Pericopsis elata* (aformosia) are internationally traded in significant volumes, including by producer countries that are members of ITTO. Not surprisingly, these proposals – which were put forward mainly by representatives of developed market economies – were strongly opposed by the producer countries affected.

Studies of previous attempts by CITES to use quantitative restrictions and bans to regulate the trade in endangered species suggest that this approach has not been an effective means of control to date (Barbier et al 1990; Burgess 1994). Problems of monitoring and enforcement are exacerbated when the producer countries affected either oppose this approach or do not receive adequate incentive (ie compensation) to participate. For example, at the 8th CITES Conference, Ghana and Cameroon opposed the proposed Appendix II listing of aformosia, and subsequently informed the conference that the timber trade was outside the scope of their countries' CITES managing authorities (ITTC 1992b). The lack of compliance by producer countries undermines the effectiveness of such policy options.

Similar problems of incentives for management, enforcement and monitoring will exist for any import quotas based on the level of 'sustainable' offtake of specific species that are imposed without the cooperation of tropical timber producing countries. For such a policy to be workable from the outset, cooperation from producer countries in determining the scientific and trade data necessary for establishing quotas for sustainable offtake and exports will be essential.

Through the cooperation of both CITES and ITTO, it might be possible to establish a comprehensive trade mechanism that establishes sustainable offtake export quotas for those species that are endangered, thus offering an incentive to both consumers and producers to accept a controlled legal trade and to enforce it.

A properly constructed trade mechanism for each endangered species, using economic instruments such as taxes and subsidies to manage trade where appropriate, could be designed to enable sufficient profits from the trade to be channelled back into producer states to encourage management efforts, and to support improved monitoring of harvesting and export activities (Burgess 1994). However, full acceptance of and adherence to any such mechanism by affected producer countries are essential to its success in terms of promoting compliance with export regulations and cooperation between exporters and regulating agencies.

In sum, the use of quantitative restrictions to regulate the trade in tropical timber products can suffer many of the same problems associated with complete and selective trade bans. Although banning or controlling the trade in specific timber species may be a necessary short term response if the species are endangered, more effective and innovative long term solutions for management of the trade are required. A more comprehensive trade mechanism that establishes sustainable offtake export quotas for endangered tropical timber species would offer a better incentive to both importers and exporters to accept a controlled legal trade and to enforce it. CITES and ITTO could work together to establish such a mechanism.[6]

Trade Taxes

As discussed in Chapter 9, an international trade tax would be a powerful means of increasing the costs associated with trading tropical timber products. Such a tax could be implemented at either the exporting or the importing end of the international tropical timber trade. In order for this tax to be effective in achieving a reduction in the amount of tropical timber products traded, it would have to be sufficiently high to stimulate producers and consumers to change their patterns of resource use. However, where the tax is implemented will effect the extent to which the cost of the tax is passed on to the producers or the consumers.

An international trade tax on tropical timber products suffers from the same sort of problems of implementation and effectiveness as those faced by trade bans and quantitative restrictions (see above).

First, it is unlikely that tropical timber exporting countries would endorse an international tax that would discriminate against tropical timber products. Chapter 7 noted that these countries have employed their own export taxes on selective products (eg log exports, and more recently sawnwood exports). However, the main purpose of such policies has not been to *reduce the overall trade* in tropical timber products, or for that matter timber-related tropical deforestation, but to *change the pattern* of their tropical timber exports to higher valued products – thus stimulating wood-based processing industries.

Second, a tropical timber trade tax would in any case probably be *ineffective* in reducing timber-related tropical deforestation – for the same reasons outlined above in the discussion of trade bans. In the short run, producer countries may increase their volumes of exports to maintain their level of foreign exchange

---

6 As an encouraging sign, at the 8th Conference, the ITTO representative noted and welcomed the call made by several (government) parties for increased cooperation between CITES and ITTO (ITTC 1992b).

earnings. Most likely, an overall trade tax might *reduce the incentives* for sustainable timber management, as the tax would lower the net returns from production and trade. As a result, the tax may actually prove to be *counter-productive* by encouraging the conversion of permanent production forests and any remaining 'virgin' forest areas to alternative uses in the long run.

It is often argued that some of the problems associated with an indiscriminately applied tax on all tropical timber products could be avoided by a tax imposed selectively on 'unsustainably' produced tropical timber products. For example, selective import duties on tropical timber products could be applied, but sustainably produced timber could be allowed in importing markets duty free. However, as argued in the case of selective bans and quantitative restrictions, there are obvious problems of certification, monitoring and enforcement that need to be sorted out – along with the political feasibility of such an approach. The issue of certification is an important one, and we discuss it in more detail below.

The economic impact of an import tax on tropical timber was analysed in a CGTM simulation for this book (see Perez-Garcia and Lippke 1993). The policy scenario simulates the tax by assuming a 10 per cent increase in the 'transfer cost' of products reaching destination countries. The results suggest that:

- Some product exports are expected to be driven out by competition with domestic supplies.
- Log prices in consumer countries rise but fall in producer countries. As a consequence, most consumer countries reduce their imports of tropical timber logs.
- With such a large impact on log imports for their processing industries, consuming country demand for processed products results in a small increase in tropical plywood exports, at least initially. However, as this import demand slackens and as log scarcity in producer countries becomes binding, tropical plywood exports fall significantly after 1995.
- All tropical hardwood suppliers suffer a decline in production, at least initially. Indonesian production declines by up to 2 mn m^3 compared with base case projections. However, the declines are not permanent in all regions. For example, Malaysia east (includes Sabah and Sarawak) experiences an initial decline of 4 mn m^3, but its log production generally exceeds the base case scenario after 1995.
- The net effect of log production declines in producing countries, distortions in trade patterns and falling producer prices for tropical timber harvesting will be less motivation to manage production forests sustainably, as well as reduced revenue to finance these investments.

In sum, a trade tax on tropical timber products would not be an effective means of encouraging sustainable production forest management, and may even be a disincentive for sustainable management.

Trade Subsidy

Unlike a tax, a trade subsidy essentially reduces the costs associated with the trade and encourages higher levels of trade in tropical timber products. It is argued that a subsidy could be used on differentiated products that are traded internationally to alter patterns of trade. For example, a subsidy could be levied on 'sustainably produced' tropical timber in order to encourage the trade in products from this timber rather than from unsustainably managed tropical forests.

The rationale for a trade subsidy applied to sustainably managed timber usually rests on the arguments that:

- sustainable management of production forests will in most instances add to the cost of logging; and,
- prices at the consumer end of the trade chain are not high enough to create enough value at the forest end of the chain to create the incentives for sustainable management, or if prices *are* high enough at the consumer end then the international distribution system leaves too little of the net returns at the forest level to encourage sustainable forest management.

These issues were recently discussed at an ITTO workshop on trade-related incentives held in Melbourne (ITTC 1992d). The report of the Melbourne Workshop argued that only 'if the trade is not and cannot be capable of financing sustainable management in both the revenue generation and distribution aspects should the matter of incentives based on subsidization come into the reckoning'. However, the workshop also concluded that there is currently very little empirical verification that the above arguments are correct. The only indisputable fact is that 'for industrial timber production to be sustainable, harvesting must have a very low impact'.

In Chapter 7 and in Barbier et al (1994), the implications of sustainable management restrictions were examined, using the example of Indonesia. A surprising result of our analysis is that, even if increases in logging costs of 25 and 50 per cent are assumed, the resulting impacts on the rest of Indonesia's forestry sector, including its processing industries and production, seem to be much less. In particular, Indonesia's sawnwood and plywood exports appear to be the least affected by the increased harvest costs, which would suggest that external demand factors exert an important counter-acting influence. The impacts of the increased harvesting costs would presumably be less severe if sustainable management is 'phased in' over a number of years. In contrast, there appears to be some direct reduction in timber-related deforestation, but even this may be outweighed by the improvement in forest management and protection resulting from qualitative changes in timber stand management practices and ownership. The Indonesian example would suggest that the need for a trade subsidy is less serious than many believe.

A major problem is that a subsidy for 'sustainably' produced timber exports could easily become a disguised means of trade promotion. International competitors would insist on verification that this policy is subsidizing sustainably managed timber only and that the additional costs of producing this timber justifies such a subsidy. This would again require careful and internationally agreed monitoring, enforcement and certification procedures. In addition, as the Indonesian example illustrates, it is not all that clear that increased harvesting costs put exporting timber-processing industries at a competitive disadvantage. Thus simply because harvesting costs may increase as a result of sustainable management practices does not necessarily justify the use of trade subsidies for exported tropical timber products. In addition, trade subsidies may in fact encourage *inefficient* resource use and have a detrimental impact on the forest base.

In sum, our analysis lends support to the view of the Melbourne Workshop that trade subsidies ought to be a last resort, and in any case, are 'no substitute for forest-related incentives' (ITTC 1992d). If the problem is one of increasing the net returns and incentives at the forest management level for sustainable production, then as discussed in Chapter 6, priority must be given to correcting domestic market and policy failures in producer countries that discriminate against such management practices. Existing trade policies in those countries that exacerbate 'unsustainable' and inefficient management should also be reviewed and corrected. Progress toward sustainable management should come first before any consideration of trade or production subsidies be entertained.

However, to the extent that additional revenue needs to gained from the tropical timber trade or through non-trade-related sources to improve the incentives for sustainable management of production forests, then mechanisms to raise this revenue ought to be explored. We discuss this issue next, focusing on the extent to which each revenue-raising measure can be used to complement the forest-related incentives for sustainable management.

RAISING REVENUES FOR SUSTAINABLE FOREST MANAGEMENT

The main rationale for providing finance for assisting tropical forest countries in moving towards sustainable forest management is that there is an important principle of *international compensation* at stake. There are essentially three reasons for this argument:

- As discussed above, it is suggested that timber exporting countries receive an insufficient share of the returns from tropical timber product exports – at least to incur the additional harvesting costs and other economic impacts of sustainable timber management.
- Implementation of the forestry policies and regulations required to ensure the proper enforcement and monitoring of sustainable management of production forests will impose substantial additional costs on producer countries that they will find difficult to afford.
- If all nations benefit from the global external benefits resulting from sustainable management of large tracts of tropical forest lands, then the international community should compensate producing nations for the loss of potential income that they would otherwise incur by tropical deforestation, timber sales and conversion of forest land to other uses.

In the discussion of trade subsidies, it was noted that the first point is difficult to substantiate. As will be discussed further below, the issue may be less to do with whether the unequal distribution of revenues along the chain of trade *reduces the incentives* for sustainable management of the forest level but whether any excess revenues along the chain can be tapped for *additional funds* to assist sustainable management of tropical production forests.

The second and third points are much more relevant to the argument for international compensation. In Chapters 5 and 9, it was argued that *international compensation* for tropical forest countries for their role in maintaining a resource that has value on a *global level* was a fundamental basis of multilateral policy action. However, the second and third points also suggest that compensation is needed by producing countries for the income they may forego in protecting their forests and for the additional costs incurred in implementing sustainable management practices for their production forests.

A policy simulation in the CGTM for this book was conducted as an indication of the additional economic impacts to tropical forest countries of 'setting aside' some of their forest resource base (see Perez-Garcia and Lippke 1993). Essentially, this was simulated by a reduced timber supply scenario where the inventory of commercial tropical hardwood resources is reduced by 10 per cent – which is equivalent to land being taken out of product forests and permanently protected. The result is that severe shortages in log production and higher saw log prices are experienced in tropical forest regions, notably in Malaysia and Indonesia. The model indicates that such reductions in supply would result in a loss of wealth for tropical timber producing countries. Over the long run, permanent set asides would mean that the remaining production forest inventory could not support as high a level of sustainable harvest as under base case projections.

There are also indications that the additional costs required to implement sustainable forestry management policies and regulations are significant. Drawing on the work by Poore et al (1989), a rough assessment of the resources needed by

Table 10.1 Estimates of the resources needed by producer countries to attain sustainability by the year 2000

Policy actions needed	Resources needed (US$ mn year)		
	Existing[a]	Additional	Total
Securing the permanent forest estate	60.2	191.3	251.5
Mapping	0.0	6.0	6.0
Inventories	7.3	28.2	36.5
Production and consumption	2.1	5.0	7.1
Land use planning	0.8	19.3	20.2
Policy and legislation	50.0	132.8	182.1
Implementing sustainable forest management	84.0	108.5	192.5
Demonstration forests	4.0	26.0	30.0
Implementation (mid-range)	80.0	82.5	162.5
Improving resource utilization	1.0	13.0	14.0
Trials and missions	1.0	13.0	14.0
Improving the social and political environment	0.0	8.1	8.1
Social sciences	0.0	4.4	4.4
Political and consumer awareness	0.0	3.7	3.7
Strategy plans and progress reports	4.0	9.2	13.2
Total	149.2	330.1	479.3

Notes [a]Existing resources are estimates of producer country expenditures based on the current status of activities undertaken, including those using external funding
Source ITTC (1992) based on Ferguson and Muñoz-Reyes Navarro (1992)

producer countries to attain sustainability by the year 2000 was conducted on behalf of ITTO (Ferguson and Muñoz-Reyes Navarro 1992). The estimates are shown in Table 10.1. Although preliminary and very approximate, they show that an additional US$ 330.1 mn is required each year in order to assist producer countries of ITTO in attaining the Year 2000 Target. Moreover, in making their estimates, the consultants suggest that the range of measures proposed is 'modest and perhaps conservative'. Thus, at best the estimates are an indication of only the minimum financing required.

Sizeable though this figure may seem, it is less than the estimated amount required for sustainable management of *all* tropical forest resources. For example, Agenda 21 of the UN Conference on Environment and Development (UNCED) has estimated that international financing of over US$ 1.5 bn annually will be required by tropical forest countries to reduce deforestation (ITTC 1992c).[7] Using these two estimates as the broad 'bounds' on the type of financing required, we suggest that additional funds required by producer countries to implement sustainable management of their tropical forest resource would be in the range of *US$ 0.3 to 1.5 bn annually*.

Although these figures would suggest the need for additional financial assistance for producer countries, the real issue that the international community must face is whether the financing ought to be raised from the tropical timber trade or from other sources. In Chapter 9, several policy options for raising revenue for sustainable management of tropical forest management were considered. These include:

- redirection of existing revenue from the trade;
- appropriation of additional revenue from the trade; and,
- additional funding from sources external to the trade.

Although these policy options are not necessarily mutually exclusive, they will be assessed in turn.

Appropriating Existing Revenue from the Trade

A recent study for ITTO conducted by the Oxford Forestry Institute (OFI) has argued the case for a *tax transfer* of revenue from the trade between consumer and producer countries (OFI 1991). The main rationale behind the transfer is that:

- Revenues from stumpage (royalties) and other taxes accruing to producer country governments from tropical timber production and trade are often low in relation to the consumer value of products; ie, producer country govern-

---

7 The proposed international financing is for, specifically: sustaining the multiple roles and functions of all types of forests, forest lands and woodlands (US$ 860 mn pa); enhancement of the protection, sustainable management and conservation of all forests, the greening of degraded areas through forest rehabilitation, afforestation, reforestation and other rehabilitation measures (US$ 460 mn pa); and promoting efficient utilization and assessment to recover the full valuation of the goods and services provided by forests, forest lands and woodlands (US$ 230 mn pa).

ments capture a low proportion of the total economic rent earned through the trade.
- Revenues from taxes on imported tropical timber products accruing to consumer governments, such as VAT, are relatively large.
- Consequently, a relatively modest reduction in the rate of taxation at the consumer end of the chain would allow a reasonably large increase in the stumpage value of the resource *without* affecting the end price of the resource.

The first column of figures in Table 10.2 illustrates the current situation. Approximately 10 to 35 per cent of the increase in value added of log and processed

Table 10.2 Effects of a tax transfer on the distribution of value added in the timber trade

	Share of total value (%)				Change relative to current share (%)
	Current		Tax transfer		
Log exports[a]					
Producer country	9.2		11.1		21
Operating cost		5.7		5.8	3
Profits		0.9		1.1	19
Gov't revenue		2.6		4.2	59
Consumer country	90.8		88.9		−2
Operating cost		40.1		41.3	3
Profits		26.2		27.5	5
Gov't revenue		24.5		20.1	−18
Totals	100.0	100.0	100.0	100.0	
Timber exports[b]					
Producer country	10.5		12.6		19
Operating cost		7.5		7.7	3
Profits		1.0		1.2	20
Gov't revenue		2.0		3.6	82
Consumer country	89.5		87.4		−2
Operating cost		40.4		41.5	3
Profits		24.1		25.1	4
Gov't revenue		25.1		20.8	−17
Totals	100.0	100.0	100.0	100.0	
Product exports[c]					
Producer country	35.3		37.1		5
Operating cost		20.1		19.3	−4
Profits		8.4		9.0	7
Gov't revenue		6.9		8.8	29
Consumer country	64.7		62.9		3
Operating cost		19.3		19.8	2
Profits		22.2		22.8	3
Gov't revenue		23.2		20.2	−13
Totals	100.0	100.0	100.0	100.0	

Notes: [a] Tropical logs exported by producer countries and all processing in consumer countries
[b] Primary processing of timber in producer countries and secondary processing in consumer countries
[c] Export of final product with all processing done in the producer country

Source: Based on Figures 4.1, 4.2 and 4.3 in OFI (1991).

products respectively is actually retained within producer countries. Of this amount, only a small percentage accrues as government revenues. At the forest stand, the royalties and other fees the producer country government charge for commercial logging are very low.[8] Typically, royalties are in the range of US$ 2–10 per m^3 felled – or approximately US$ 10–300 per ha according to the amount removed per unit area.

Thus the stumpage value of the timber, as reflected by these royalties, is only a small proportion of the final value of the product. For example, conversion losses of logs to processed products can result in around 3 m^3 of logs being harvested to produce 1 m^3 of final product. The total stumpage value collected by producer country governments of US$ 6–30 per 1 m^3 of end product often represents less than 1 per cent of the final value of the product being sold in consumer countries. In contrast, Table 10.2 shows that consumer country governments capture approximately 25 per cent of the final value of tropical timber products as revenue.

Table 10.2 also indicates how a tax transfer can improve this situation. In the current scenario, producer country governments are assumed to collect a royalty of US$ 9 per m^3, and consumer country governments impose a VAT on final products of 15 per cent. However, in the tax transfer scenario, VAT could be reduced to 7.5 per cent and the royalty increased to US$ 30 per m^3 – and the final price would be left unchanged. Depending on the product, tax revenues of the consumer country governments would fall by 13–18 per cent. These losses are offset by important incentives:

- Due to the increased royalties, producer country government revenues rise dramatically, by 29–82 per cent, which would provide additional resources for sustainable management expenditures (see Table 10.1).
- The increased royalty would also raise the sale price of logs within producer countries, which could help to control over-exploitation of timber and inefficient use of logs in processing (see Chapter 6).
- Despite the increase in the price of logs, within the forestry sector of producer countries profits would still rise modestly. A slight rise in profits from the trade would also occur in the consumer countries.

In short, the main implications of a tax transfer would be to change significantly the *distribution* of economic rents from the timber trade as an incentive for sustainable management in producer countries. The main advantage would be that additional funds for this purpose could be raised *with little or no* effect on final product prices. As indicated in Table 10.3, just under US$ 1.5 bn in additional funds could be raised by producer countries through this means – closer to the

---

8   The timber royalty essentially approximates the *stumpage value* of the commercial timber stand – ie what the government charges for the standing tree. This value is also the residual after deducting all operating costs, commercial profits and taxes from the market price of logs or products. Other forest stand fees (eg for concession licence, felling, reforestation etc) in addition to the royalty can also be considered part of the residual, or stumpage, value of harvested timber. Thus, as discussed in Chapter 6, stumpage value is equivalent to the *economic rent* of timber harvesting, and as a result should reflect the long run scarcity value of timber.

Table 10.3 Revenue transfers available from value added in the tropical timber trade, 1990 (US$ mn)

	Log exports	Sawnwood exports	Processed product exports	Total
fob export value[a]	2,280	2,150	4,150	8,580
Final end use value[b]	46,588	29,189	12,858	88,635
Tax Transfer Scenario[c]				
Consumer country				
Current government revenue	11,414	7,326	2,983	21,724
Post-transfer revenue	9,364	6,071	2,597	18,033
Revenue lost/transferred	2,050	1,255	386	3,691
Producer country				
Current government revenue	1,211	584	887	2,682
Post-transfer revenue	1,957	1,051	1,132	4,139
Revenue gained/transferred	745	467	244	1,457

Notes: [a] Based on FAO (1992)
[b] Based on ratios of FOB value to final product value in OFI (1991)
[c] Based on Table 10.2
Sources: FAO (1992); OFI (1991)

'upper bound' of the estimated financing for sustainable management required by these countries.

However, consumer countries are likely to be concerned about the fiscal and political implications of a tax transfer scheme. First, as Table 10.3 shows, consumer country governments would have to forego over US$ 3.6 bn in tax revenues from the trade – more than 2.5 times what producer country governments gain in increased revenues. This implies a substantial net loss in revenue 'captured' from the trade. Second, exempting tropical timber products from VAT or other taxes could prove politically problematic in that it could set the precedent for other goods being exempted from taxes on 'environmental' grounds. Forest industries in temperate forest countries and their governments could also press for similar treatment on the same grounds – the need for additional investment for sustainable management or for compensation for past investment.

Finally, consumer countries would insist on an internationally agreed monitoring and enforcement system that would ensure that:

- producer countries did respond by raising royalties; and,
- the additional revenue raised was spent on sustainable forest management.

On the other hand, producer countries would consider too much external supervision to be 'interference' with their internal affairs and sovereign rights over their resource base. Carefully negotiated bilateral and multilateral agreements would be necessary to avoid confrontations and to make the system workable.

One possible variant on the tax transfer scheme would be a *revenue transfer* scheme. Rather than lower their VAT or other taxes on the tropical timber trade, consumer country governments could instead transfer directly some proportion, or all, of the revenue raised through these taxes to producer countries. Although the consumer country governments would still forego substantial revenues, they could ensure more direct control, and thus leverage, over the allocation of funds to sustainable management. For example, they could require that producer countries demonstrate substantive progress in implementing sustainable timber management – such as formally adopting and taking steps to implement ITTO's Year 2000 Target – as a condition for receiving the rebated revenue from the trade. In addition, consumer country governments would most likely have to forego *less* revenue if it were directly transferred to producer countries to help them meet their US$ 0.3–1.5 bn target then under the tax transfer scheme (see Table 10.3). As indicated in the table, due to 'leakages' to other sectors in the tropical timber trade, a tax transfer scheme would require consumer country governments to forego around 2.5 times more tax revenue in order to allow producer country governments to achieve their target.

However, a revenue transfer scheme would still require an internationally agreed monitoring and enforcement system. Producer country governments would probably be more concerned about the direct control and conditions that consumer country governments could assert over a revenue transfer scheme. Producers would insist on full involvement in the establishment, implementation and monitoring of any such scheme.

In sum, both revenue and tax transfer schemes are similar means of appropriating existing funds from the tropical timber trade for sustainable management. Both have the attraction that they would allow substantial funds to be raised for producer country investments in sustainable forest management *without* affecting product prices significantly. However, consumer countries are more likely to favour a revenue transfer scheme, whereas producers may prefer the tax transfer scheme. Both would require international cooperation and agreement over monitoring and enforcement.

Appropriating Additional Revenue

One of the main drawbacks to the tax or revenue transfer scheme is that consumer country governments may be reluctant to see tax revenue transferred to other countries. Although the sums involved may be small compared to the overall government budgets of many consumer countries, current concerns over budget deficits in developed market economies may mean that governments are reluctant to lose any tax revenue. Thus in recent years there has been renewed interest in the use of trade instruments in appropriating additional tax revenue from the trade for sustainable forest management.

One study by the NEI has indicated that a 1–3 per cent surcharge on the tropical timber imports of the EEC, Japan and USA would raise approximately US$ 31.4 to 94.1 mn with little additional distortionary effects (NEI 1989). If endorsed by the ITTO and sanctioned through the ITTA, the import surcharge would be within GATT rules, again through Sub-article XX(h). A differentiated surcharge could also be imposed so that imports of processed tropical hardwood products face less discrimination than logs, thus reducing existing distortions from escalating tariffs. As suggested by the NEI study, the funds raised could be transferred to the ITTO for distribution, possibly through specific projects and programmes. Other forms of collecting additional funds were found by the study to be less desirable.

For example, imposition of an export levy by producing countries themselves has the advantage of directly addressing the forest management systems of those countries, but presents obvious problems of monitoring and evaluating success in achieving sustainable management. If the funds were transferred to ITTO, transaction costs could be high, the same rate would need to be implemented in all producer countries simultaneously, and fewer funds would be raised than through an import surcharge. A parafiscal tax on all timber sold in consumer countries has the advantages of taxing all kinds of timber equally, including non-tropical products, and of generating large revenues at a low rate of taxation. However, such a tax has complications for external trade policies (eg temperate forest countries could claim unfair discrimination) and 'harmonization' of internal tax rates within trading blocs (eg the EEC's Internal Market Strategy). Finally, a voluntary surcharge collected by the tropical timber trade itself could raise funds transferable to ITTO without requiring additional national legislation. Unfortunately, compliance and effectiveness may be diminished because of:

- the current over-capacity and low net profit margins faced by many international traders and wood processing industries;
- the lack of effective control measures to ensure collection with equal efficiency in all consumer countries; and,
- the possibility of entry and exit by traders in the industry to avoid payment (NEI 1989).

Buongiorno and Manurung (1992) also examine the scope for a 5 per cent revenue-raising import levy on tropical timber by the UCBT of the EEC. The results indicate that tropical timber exporters would lose around US$ 44.8 mn in trade earnings, with Indonesia and Malaysia suffering the worst, but importing countries would earn US$ 87.7 mn in additional revenues. Thus, if the funds raised by the tax were rebated to exporting countries, they could be made better off by over US$ 40 mn. Moreover, a tax by UCBT countries would have little effect on total world trade of tropical timber products. The model shows that tropical timber exporting countries would compensate for any declines in imports by UCBT countries through increasing exports to non-UCBT countries.

One of the major concerns of producer countries is that any revenue-raising import surcharge, even at very low levels, would be distortionary. In particular, if the tax was levied by all importers, then there could be a more significant impact on total world trade of tropical timber products. Moreover, such a tax would provide unfair protection to temperate forest based industries of the developed market economies. For example, Buongiorno and Manurung suggest that producer countries may prefer a tax on exports rather than an import surcharge by UCBT countries. This would give producers more direct control over the proceeds of the tax. In addition, an export tax would affect all import markets rather than just one, thus spreading the costs of sustainable management to all producers and consumers.

To examine the impacts of an import surcharge on the export markets of a producer country, we conducted two policy scenarios involving a 1 and 5 per cent surcharge respectively on Indonesia's timber trade (see Barbier et al 1994). In the case of a 1 per cent surcharge, a total of US$ 23.1 mn (1980 prices) in revenue would be raised, with US$ 5.8 mn and US$ 17.3 mn from Indonesian sawnwood and plywood exports respectively. For the 5 per cent surcharge, a total of US$ 113.9 mn (1980 prices) in revenue would be raised, with US$ 28.5 mn and US$ 85.4 mn from Indonesian sawnwood and plywood exports respectively. When compared to the revenue estimates from the NEI (1989) study, the above figures suggest that, in the case of Indonesia, applying the import surcharge to plywood would significantly raise the total amount of financing appropriated.

The results from Barbier et al (1994) also confirm that a small import surcharge would have very little distortionary effects on Indonesia's timber product flows and prices. There would also be a negligible direct impact on deforestation. However, around the 5 per cent level, the impacts of the surcharge on exports in particular would become more noticeable. Thus from the standpoint of minimizing additional distortionary effects, the policy scenario confirms that an import surcharge of less than 5 per cent would be optimal.

A more pertinent issue is whether it is worth imposing an import surcharge to raise revenue for sustainable management of tropical forests. In the simulation model, the same result, and thus the same amount of revenue, could be achieved if Indonesia raised the money for its own 'sustainable management' initiatives by imposing a revenue-raising export surcharge of 1–5 per cent.

In sum, it may be possible to raise revenue through an import surcharge of less than 5 per cent with minimal distortionary effects on the tropical timber trade and production. The key issue is whether this is the most efficient and appropriate means of raising finance for improved sustainable management of production forests.

The imposition of an export levy by producing countries themselves has the advantage of directly addressing the forest management systems of these countries, as well as avoiding the transaction costs involved in international transfers, but prevents obvious problems of monitoring and evaluating success in achieving sustainable management. The counter-argument is that an import surcharge not only has problems of transaction costs and administration, but that it is also possibly discriminatory if it is limited only to the *tropical* timber trade. Moreover, the import surcharge-cum-international transfer mechanism would still require the cooperation of producer countries, as well as raise similar problems of monitoring and evaluation of progress towards sustainable management and expenditures.

Finally, there is the issue of whether the amount of funds raised through any trade surcharge would be adequate for the task, and whether it would be more appropriate to raise additional large-scale funding *outside* the timber trade altogether. The studies undertaken so far suggest that the amount of *net* funds raised from a trade surcharge of 1–5 per cent may fall short of the approximate target of US$ 0.3–1.5 bn required annually by producer countries as additional resources for sustainable management of their forest resources.

Funding from outside to the Trade

There is also a strong rationale for additional funds to be made available to producer countries for sustainable forest management from sources outside the tropical timber trade. Comprehensive international agreements, targeted financial aid flows and compensation mechanisms to deal with the overall problem of tropical deforestation may ultimately eliminate the need to consider intervention in the timber trade. Given that commercial logging is not the primary cause of tropical deforestation, such approaches would avoid unnecessary, and possibly inappropriate, discrimination against the timber trade. On the other hand, a comprehensive tropical forest agreement seems much more difficult to negotiate and raises its own problems of workability and effectiveness. [9]

It had been hoped that the US$ 1.5 bn annual target for additional funds for sustainable forest management in Agenda 21 would be the basis for an international Forestry Convention to be ratified at UNCED in June 1992. Unfortunately, only a statement of Forest Principles was agreed at UNCED, without formal pledges of increased financial assistance. With the deterioration in international economic conditions and increased financial insecurity since the conference, we have seen developed market economies cut back, rather than increase, their commitments to assisting developing countries. Already, some promises made at UNCED seem destined to be unfulfilled. It is unlikely in this economic climate that a concerted international effort to increase substantially bilateral or multilateral aid flows for sustainable production forest management would be successful, given the rapidly diminishing aid budgets of the developed economies.

Nevertheless, there still remains the possibility of designing new sources of financial assistance that are separate from existing developing country aid budgets. The Forestry Principles are effectively a stepping stone in that direction, and international commitments through the Tropical Forest Action Plan (TFAP) and Global Environmental Facility (GEF) continue to reinforce the global interest in forestry and biodiversity protection. A case could be made for more comprehensive international agreements to raise revenues for sustainable management of tropical forests, including production forests. The main focal points of such agreements should be the development of revenue-raising mechanisms other than trade interventions or reliance on existing aid budgets.

For example, Amelung (1993) has explored some of the conditions necessary to ensure efficient international compensation payments, arguing that payments are a better solution to controlling tropical deforestation than trade barriers as they offer the chance to 'internalize' global externalities and improve incentives for management. Moreover, rather than focus exclusively on interventions in the timber trade to finance compensation, Amelung (1993) opts for the establishment of an international rainforest fund, as proposed by UNEP, in order to avoid free-riding among non-tropical countries. To be effective, compensation payments should meet the following criteria:

- To avoid principle-agent problems, ie contract default and cheating by tropical forest countries, payments should be made periodically rather than as once-and-for-all lump sum transfers. Periodic payments would also facilitate renegotiations, if required.
- Payments should be tied to conditionalities with respect to protecting tropical

---

9 See Barrett (1990) for a general discussion of the difficulties involved in securing international environmental agreements.

forest reserves, reform of domestic market and policy failures affecting tropical forests, and generally 'foster economic activities that do not destroy the regenerative capacity of this ecosystem and the development of the necessary technologies', which presumably includes sustainably managed timber production.

- Contributions by non-tropical countries should be tied to the gross national product (GNP) of these countries. As all people are affected more or less equally by the global externalities of forest clearing, countries with a larger population are made less well off. In addition, countries with higher per capita GNP tend to have lower marginal utilities of consumption and thus lower discount rates. Since the preservation of tropical forests provides long-term benefits, wealthier nations should have a higher consumer preference for their protection.

Similar arguments are put forward by Sedjo, Bowes and Wiseman (1991), although their preference is for the establishment of a global system of marketable forest protection and management obligations (FPMOs). Under a voluntary global forestry agreement, FPMOs would be distributed to all signatories, probably through criteria based on a mix of GNP levels and forest area. Holders of FPMOs must either: i) fulfil the obligation 'on the ground'; or, ii) induce another agent to assume the obligation, presumably in exchange for a payment. Thus, countries with large obligations and small forests would have 'excess' obligations and hence be forced to meet these externally. Countries with large forests and small obligations could meet them internally but would then have an 'excess' of forests. They could then negotiate with countries that have 'excess' obligations.

The advantages of such a system are that, while not requiring compliance, it provides incentives for nations to comply out of their own self-interest, as well as achieving the objectives of international compensation. It would ensure that non-tropical forest countries (eg mainly industrialized countries) would bear a substantial portion of the costs. Also, it would not excessively limit production of traditional and commercial forest products. The disadvantages would be the high degree of monitoring required, the need for a 'clearing house' for international trade in permits, and the difficulties of negotiating a comprehensive international agreement to establish the system. On the last point, the authors argue for the establishment of a 'transition' system with only a few major industrialized and tropical forest countries involved, and initially limited to mainly bilateral agreements with little trading.

In sum, it seems unlikely that in the near future producer countries will be able to obtain additional funds for sustainable forest management from traditional aid sources. In cooperation with other international fora, a recognized body such as ITTO could take the lead in the post-UNCED era to promote international negotiations leading towards an agreement for new revenue-raising mechanisms for sustainable management of tropical forests, including production forests. Although such negotiations are difficult and arduous, they may be the best hope of ensuring a global commitment to controlling tropical deforestation, as well as the sustainable management of production forests.

CERTIFICATION

Many of the trade policy options discussed above were concerned with using some form of certification as a means to distinguish between 'sustainably' and 'unsustainably' produced tropical timber products. A comprehensive and internationally agreed certification scheme therefore seems to be a *necessary complement* to trade policy options for sustainable forest management. However, the term *certification* has been used variously to mean:

- *Product labelling* – labelling all products that include tropical wood with a label indicating whether or not it is sustainably produced.
- *Concession certification* – certifying all the timber produced by a specific concession to be 'sustainably produced'.[10]
- *Country certification* – certifying all timber products from a country that can prove it has begun complying with an internationally agreed sustainability objective, such as ITTO's Year 2000 Target and its sustainable management guidelines.

Product labelling

In our opinion, mandatory product labelling – even if it is internationally agreed – is the most difficult of the certification procedures to implement and verify. Given the vast array of tropical timber products traded, and the degree of processing involved in producing final products, it will be extremely difficult to set up a workable product labelling scheme that will be internationally verifiable. It is made more problematic by the fact that the end uses of tropical timber are often not discrete products, but components of products and composites, as well as basic structures, fixtures and fittings (see Chapter 4). The real danger is that any such scheme of product labelling will essentially work as a powerful non-tariff barrier to discriminate against the import of tropical timber in general. This will particularly be the case if it is implemented unilaterally or on a piecemeal basis, ie by individual importers or by selective trading blocs. This study has demonstrated that further restrictions on the trade in tropical timber products will not be effective in reducing tropical deforestation, and in fact may accelerate it.

Concession certication

Concession certification essentially involves:

- assessment of a forest concession in terms of compliance with sustainable management guidelines;
- monitoring of actual forestry practices in the concession, including volumes sold and destination, up to the retail level; and,
- ensuring that each product produced with timber from that concession has appropriate certification to verify its origin.

---

10 A close alternative to concession certification would be company certification.

In theory, the concession certification approach appears to offer the best means of guaranteeing that the tropical timber trade is committed to securing timber from only 'sustainably' managed sources. It has the possible attraction to traders and timber companies of earning 'brand recognition' for buying timber supplies from sustainably managed concessions. Promotion of this fact could be an important 'selling point' as part of the promotion of tropical timber products in consumer markets – much like current campaigns to sell 'organically grown' and 'environmentally friendly' products. These positive features of concession (or company) certification suggest that there are excellent incentives for 'sustainably managed' concessions and companies to promote their products through a voluntary labelling scheme. A group of countries, and even their host producer countries, may even develop a common 'voluntary' labelling scheme. Such efforts are essentially elements of a good marketing and export promotion strategy. However, in contrast, it will be difficult to make a comprehensive and internationally agreed concession certification scheme, that is *mandatorily* imposed on all concessionaires of a producer country, workable and effective.

First, there are extremely high costs of monitoring, enforcement and verification. Such a scheme would therefore require substantial funding on an international level. More importantly, concession certification would add substantially to the additional financing already required to implement sustainable forest management (see the previous section). As discussed above, the questions of who pays these additional costs, how the money ought to be raised, and how to implement such a mechanism are not easy to resolve – especially when not all concessionaires in producer countries may be willing to accept a mandatory scheme.

Second, producer countries and companies may object to detailed monitoring of all aspects of their forest industry production, from harvesting to processing to trade. It is unlikely that a team of visiting international inspectors would be able to monitor the 'wood chain' starting at the concession level in all producer countries on their own. In-country inspectors (such as in the CITES system) would probably also be required, but there will inevitably be conflicts over their level and source of funding and over their degree of autonomy.

Third, concession or company certification in itself does not involve support for forest management administration and services of producer countries. If anything, a comprehensive mandatory scheme may impose additional costs on forest departments. Unless adequate compensation is provided, the incentives to support or enforce the scheme by forest personnel may be lacking.

Fourth, concession-level monitoring also requires close monitoring of products at the retail end of the trade. For consumer products made solely from one type of wood from a single source, verification would be fairly straightforward. However, for composite products, products containing one or more different tropical timber components, or for tropical timber used as part of basic structures, fixtures and fittings, the process will be more difficult. The additional costs incurred in acquiring tropical timber with the appropriate certification may discourage its use compared to alternative materials in end use markets (see Chapter 4). Again, it is possible that such a certification scheme will:

- act as a restriction on trade and use of tropical timber for many final end uses; and,
- increase the incentives to find 'loopholes' or cheat the scheme (eg obtaining similar timber from 'unverified' sources in the same country; concessions 're-selling' timber from other sources; forging of certificates).

Country certification

Given the difficulties with product labelling and concession certification, we feel that the most appropriate scheme at least initially would be *country certification* tied to an international commitment by both producer and consumer countries to adopt policies and practices towards encouraging sustainable management of production forests and timber products, such as outlined in ITTO's Year 2000 strategy. The main point of such a scheme would be to ensure that a producer country is implementing policies, regulations and management plans that ensure substantial progress towards achieving a sustainability strategy. In return, the tropical timber products of that country will be certified as coming from a 'Target 2000' country, which should give them easier access to import markets in developed economies.[11]

There are several reasons why country certification may be more effective and workable:

Cheaper and easier

Certification at the country level will be less costly and more easy to implement compared to other certification schemes. Periodic inspection tours by internationally certified teams, monitoring at custom ports, reviews of forest policy and management plans would probably be sufficient for such a scheme to be effective.

Politically more acceptable

Producer countries would find country certification more acceptable politically, provided that:

- Under international auspices producer countries could help determine the certification scheme as well as any verification process.
- The certificate of origin could then be issued by the exporting country, or companies authorized by that country.
- As it is effectively a *national* sustainable management plan that is being certified, it would be up to the producer country to address adequately the problems posed by production from conversion forests, plantations, reafforestation and so forth, but once this plan is internationally verified, *all* timber products from *all* types of forests in the country would be certified.
- Development of a national sustainable management forest policy and land use plan would in turn support efforts by companies and concessionaires to develop more sustainable forest management practices, which they may then

---

11 A similar strategy has been proposed by the German Timber Trade Federation; see GHK/HDH/VDH 1992.

choose to promote voluntarily through internationally accredited product or concession/company labelling.
- It would be up to the producer country to ensure compliance with the sustainable management plan, and in cooperation with 'independent inspection', the relevant forest authorities would have primary responsibility for monitoring operations at the concession and industry level.
- In exchange for adopting the scheme and being certified, an exporting country would hope to receive improved access to international markets for their 'sustainably managed' products and, hopefully, international financial assistance to implement their sustainable management plans. This would provide an incentive for producer countries to adopt this policy approach, and hopefully to seek international accrediting for their sustainable management practices at the concession and company level.

More feasible
Consumer countries may also find country certification more feasible to implement as:

- Under international auspices, such as ITTO, consumer countries could help determine the certification scheme as well as any verification process.
- Individual timber trade products would not have to be certified – all trade products from a certified country could be safely imported, and any inspection could be conducted as routine port of entry (ie customs) procedures.
- Consumer countries would be assured that country certification would require: a policy commitment by producer country governments to manage their production forests sustainably under the ITTO guidelines and Target 2000; viable national plans to implement this policy; and a mandate to correct domestic market and policy failures that encourage timber-related deforestation.
- It would be easier to target bilateral and multilateral financial assistance for sustainable forest management – ie such flows could now be conditional on producer countries complying with the certification scheme.

One important aspect of country certification is that it does not exclude the development of other forms of certification, such as *voluntary* concession or company certification; rather, country certification may herald the beginning of increased cooperative efforts between producer countries, consumer countries, timber traders and independent monitors to develop more internationally recognized, transparent and detailed criteria for establishing and evaluating forest management practices. As it may be easier to negotiate and implement, country certification of tropical timber producers may not only lead to the development of more comprehensive international certification agreements but it would also 'buy time' for the development of similar accredited schemes for temperate timber and even plantation products.

Disadvantages
On the negative side, country certification will not appease those environmentalists and consumers that want absolute assurances that every timber product comes from a production forest that is guaranteed to be 'sustainably managed' according to strict internationally agreed rules. For many in this group, the aim is that one

day soon we will be able to agree, evaluate, accredit and monitor set international sustainable management rules for not only tropical forest production but also temperate and plantation forest production. In their eyes, country certification for tropical timber is not only a compromise on the path to universal sustainable forest management but also a potentially damaging diversion from the path of 'true' sustainable forest management.

There is an element of truth in this argument. If poorly implemented without sufficient international transparency, recognition or commitment, any country certification scheme could have little impact on improving sustainable management practices in tropical forest countries. On the other hand, as outlined above, a properly implemented scheme could actually provide the trade-related incentives needed for ensuring sustainable management of tropical production forests, which in turn would foster further internationally accredited concession, company and even product certification schemes. Although we would also like to see one day sustainable management of all production forests – tropical and temperate, natural and plantation – as outlined throughout this book the need for trade-related incentives for sustainable tropical forest production is urgent and needed now.

Summary

In sum, an internationally agreed certification scheme ought to be used as a means to facilitate trade in sustainably produced tropical timber – not as a means to restrict trade. Mandatory product labelling and concession certification are too cumbersome and restrictive to assist trade-related incentives for sustainable management, and may also fail to gain adequate producer country cooperation. On the other hand, country certification would require the active participation and verification of producer countries. Moreover, a country certification scheme may also support efforts by companies and concessionaires to develop more sustainable forest management practices, which they may then choose to promote voluntarily through internationally accredited product or concession/company labelling. In exchange, producer countries could qualify for additional financial assistance to implement sustainable management plans, as well as be assured of better market access for their exports.

CONCLUSIONS

This chapter has provided a brief assessment of the various trade policy options identified in Chapter 9. A summary of the timber trade policy options, and their advantages and disadvantages in promoting sustainable tropical forest management, is provided in Table 10.4. The overall purpose of any preferred option should not be to penalize timber production and trade but to ensure that trade works properly to reinforce the positive incentives for sustainable timber management. The best way to ensure this is to have the trade in tropical timber products work to *increase*, rather than *reduce*, the net returns to sustainable timber management. However, trade policy options are by themselves no substitute for producer countries adopting appropriate domestic policies and regulations for

Table 10.4 Timber trade policies to promote sustainable tropical forest management

Policy options	Advantages	Disadvantages
Trade restraint Import and export bans, quotas, tariffs and non-tariff barriers are used to restrict or regulate trade in forest products. Policy options include: export bans or preferential taxes to encourage value-added processing; selective import bans; punitive tariffs on 'unsustainably produced' timber; limits on trade in endangered tree species; national import quotas based on estimates of sustainable timber yield.	• Relatively simple to administer and enforce (less so for selective restraints). • Tariffs and taxes can raise revenue for government. • Selective import restraints may encourage improved forest management, by penalizing 'unsustainably produced' products in export markets. • Selective export restraints may stimulate value-added processing and boost employment in producer countries.	• May be considered discriminatory and protectionist. • Potential trade diversion to export markets which impose fewer environmental or other restraints. • Limited impact due to small scale of international trade in tropical timber. • Decrease in export prices may reduce economic returns to timber production and thus undermine incentives for investment in forest management. • Inefficient wood recovery in protected processing industry may imply increased log use and deforestation.
Trade stimulus Tropical timber importers have reduced tariffs under agreements to liberalize trade (eg the GATT), but some quotas and non-tariff barriers remain. Producer countries call for removal of import barriers, especially on higher processed products. Others suggest selective trade liberalization or targeted subsidies to promote trade in 'sustainably produced' timber and/or to compensate producers for the higher costs of sustainable forest management.	• Relatively simple to administer and enforce (less so for selective stimulus). • Consistent with the aims of global trade negotiations (eg the GATT). • Increased export prices may enhance economic incentives to invest in production forest management. • Selective liberalization or subsidies for 'sustainably produced' products may influence industry or governments to adopt better management practices.	• May stimulate unsustainable logging if not supported by appropriate domestic forestry policies (eg royalty and concession terms). • Liberalization may not effect prices due to competition from temperate timber and non-timber substitutes. • Subsidies for 'sustainably produced' products may be costly and ineffective if not strictly targeted.

Assessment of Trade Policy Options

Revenue mechanisms International financial assistance is required to support moves towards sustainable management of tropical production forests. Proposals include: redirection of existing trade revenues, eg tax relief in consumer countries linked to increased royalties in producer countries; incremental trade taxes to finance forest management; and additional funding from external sources (eg development assistance, compensatory transfer, forest offsets).

- Additional revenue may defray the costs of adopting sustainable management or expanding set asides.
- Increased revenues may help to finance government forest service activities or technology transfer.
- Increased royalties may reduce over-harvesting and improve wood recovery rates in processing.

- May be considered discriminatory and protectionist.
- Additional taxes may distort trade patterns.
- Consumer country governments may be reluctant to relinquish tax revenue raised on the trade to provide additional financial transfers to producers.
- Administration and monitoring of funding mechanisms may be contentious.
- International agreement to new financial transfer mechanisms may be difficult to negotiate.

Product differentiation Many trade policy options require a means to distinguish between 'sustainably' and 'unsustainably' produced timber. Proposals include product labelling, in some cases combined with certification of compliance with sustainability criteria. Certification may apply to forest compartments or to entire nations, or some intermediate level.

- Certification may identify forest operators, regions or nations adopting 'sustainable' management practices, offering competitive advantage or ensuring eligibility for statutory preferences.
- Labelling may enable traders and consumers to choose between more or less 'sustainably' produced timber products.

- Certification may be costly, especially at a forest concession or firm level.
- Competing labelling and certification schemes may confuse both producers and consumers.
- Schemes may be resisted or ignored if they are considered intrusive, biased or unreliable.

sustainable management of their production forests. The international community should therefore look to trade policy options that help reinforce the commitment by producer countries towards implementing a sustainable management goal – such as ITTO's commitment to the Year 2000 objective – rather than to those options that tend to work against this goal. The main conclusions of our assessment of trade policy options are:

- There is very little evidence to suggest that the option of 'doing nothing' will encourage greater trade-related incentives for sustainable management of tropical production forests, or is the appropriate response given current trends in tropical deforestation and trade.
- The problem with restrictive trade policy interventions, such as bans, taxes and quantitative restrictions, is that if they are applied indiscriminately across all tropical timber products, they will reduce rather than increase the incentives for sustainable timber management – and may prove actually to increase overall tropical deforestation. 'Selective' interventions, such as those that distinguish between 'sustainably' and 'unsustainably' produced timber, would also work to restrict trade and thus should also be avoided.
- Banning or controlling the trade in specific timber species may be a necessary short-term response if the species are endangered, but more effective and innovative long-term solutions for management of the trade are required. A more comprehensive trade mechanism that establishes sustainable offtake export quotas for endangered tropical timber species would offer a better incentive to both importers and exporters to accept a controlled legal trade and to enforce it.
- Trade subsidies to support sustainable management goals would also probably be distortionary and could lead to problems of disguised protectionism and encourage inefficiency in the forest sector. Progress toward sustainable management in producer countries should come first before any consideration of trade or production subsidies.
- General liberalization of the trade in tropical timber products is probably not feasible. However, producer countries that show demonstrable progress towards achieving an internationally agreed sustainable management objective, such as Target 2000, could be allowed better access to importing markets through the removal of specific tariff and non-tariff barriers on a case-by-case basis. Similarly, in demonstrating compliance with Target 2000, producer countries should be asked to review the implications of their own tropical timber export policies and restrictions for both timber-related tropical deforestation and international markets.
- However, there is evidence that producer countries will require additional financing of around US$ 0.3 to 1.5 bn annually to implement sustainable management plans for their tropical forests, including production forests. Moreover, the international community led by the developed market economies ought to provide this assistance, given the global values attributed to tropical forests. In the current economic and political climate, new forms of assistance other than traditional bilateral and multilateral aid flows will probably be required.

- A *revenue-raising surcharge* could be levied at either the exporting or importing end of the trade. Trade distortions would be minimized if the surcharge was kept to under 5 per cent of the value of the trade. Unfortunately, the net funds raised may be well below the annual target of US$ 0.3–1.5 bn. However, the trade should not be expected to pick up the total tab for financing sustainable management of tropical forests, but rather make a contribution towards it.
- If additional revenue were to be raised from the trade, the distribution of existing funds appropriated by consumer country governments, either through a *tax* or *revenue* transfer, would be feasible. Consumer countries would probably prefer the latter, and producer countries the former.
- In the medium and long term, more comprehensive mechanisms and international agreements for raising revenue *external* to the trade to combat the *overall* problem of tropical deforestation will have to be examined. Possible mechanisms include the establishment of a tropical forest fund or a global system of marketable forest protection and management obligations.
- Finally, any promotion of trade-related incentives for sustainable management will have to consider an appropriate *certification* scheme. Product labelling and concession or company certification are too cumbersome and restrictive to assist trade-related incentives for sustainable management, and may also fail to gain adequate producer country cooperation. On the other hand, country certification would require the active participation and verification of producer countries and satisfy consumer demand for a sustainably managed supply of timber products. In exchange, producer countries could qualify for additional financial assistance to implement sustainable management plans, as well as be assured of better market access for their exports.

In conclusion, throughout this book we have emphasized that the key factor in reducing timber-related tropical deforestation is ensuring proper economic *incentives* for efficient and sustainable management of tropical production forests. Appropriate forest management policies and regulations within producer countries ought to provide these incentives so that the *long run* income-generating potential of harvesting timber is maximized, and any significant external environmental costs associated with timber harvesting are 'internalized'. Unfortunately, as discussed in Chapter 6, domestic economic and forestry policies in producer countries have tended to distort the incentives for sustainable management of timber resources and have failed to curb the wider environmental problems associated with timber extraction. Trade policy distortions in producer and consumer countries have, if anything, exacerbated this situation.

However, by adding value to forestry options, the trade in tropical timber products *could* act as an incentive to sustainable management of production forests – provided that the appropriate domestic forest management policies and regulations are also implemented by producer countries. Nevertheless, as this chapter has made clear, many proposed trade policy interventions – such as bans, taxes and quantitative restrictions – will actually work to *restrict* the trade in tropical timber products. Such interventions will reduce rather than increase the incentives for sustainable timber management – and may increase overall tropical deforestation.

Thus our analysis casts doubt on the effectiveness of trade policy options generally in reducing tropical deforestation. Trade measures have their most direct impact on cross-border product flows and prices. Our study suggests that changes in these international flows may have little influence on the main proximate causes of tropical deforestation in producer countries. Policy interventions that restrict the trade in tropical timber products should not be implemented, as they will not only be ineffective but may also be counter-productive in reducing deforestation.

On the other hand, there does appear to be a role for tropical timber trade policies in fostering *trade-related incentives* for sustainable management. Trade policies will be the most effective if:

- they are employed in conjunction with and complement improved domestic policies and regulations for sustainable forest management within producer countries;
- they improve rather than restrict access to import markets of tropical timber products so as to ensure maximum value added for sustainably produced tropical timber exports; and,
- they assist producer countries in obtaining the additional financial resources required to implement comprehensive, national plans for sustainable management of tropical production forests.

In this final chapter we have presented the elements of such a strategy, which we believe is supported by the analysis and evidence presented throughout this book.

EPILOGUE

Since we completed our study for this book in May 1993, debate over many key tropical timber trade policy issues – most notably international certification – has intensified rather than lessened.

The Uruguay Round of the GATT has been completed – although preliminary indications suggest that the implications for the trade in tropical products may be negligible given the low tariff barriers already existing for these products. A better understanding of the implications will depend on the results of an analysis to be launched by the FAO and GATT.

Perhaps even more significant has been the continuing development of an independent international certification scheme for sustainable forest management. Although many certifying consultants and organizations have been established in recent years, a major impetus has been the creation of the Forest Stewardship Council (FSC). Formally launched in October 1993, the FSC is an international NGO that aims to act as 'an international body to accredit certifying organizations in order to guarantee the authenticity of their claims' (FSC 1993). Essentially, the FSC hopes to act as a 'certifier of the certifiers' in order to ensure that there is an internationally consistent, and above all independent, verification of the sustainability of forest management practices worldwide. To this end, principles and criteria for management of both plantation and natural forest managment have been drafted, which are intended to be equally applicable to all forests: temperate, tropical and boreal (FSC 1993 and Synnott 1993). It is hoped that the first certifying companies and organizations will be accredited by the FSC by the end of 1994 and that the first timber products from certified markets will reach consumer countries soon thereafter.

It is obviously too soon to judge how successful the FSC is likely to be. As discussed in Chapter 10, the key to the success of any international certification scheme involving tropical timber products will be the cooperation of producer countries. While increasingly resigned to the inevitability of international certification, producer countries are somewhat wary of schemes that appear to be initiated and dominated by environmental NGOs and aimed primarily at reassuring consumers in European and North American markets. In particular, concern has been expressed about the continuing use of unilateral certification and 'green labelling' schemes in some consumer markets and their effects on the trade (*INER* 1994 and *MTIDC News* 1994).

Because of their fear over the potentially discriminatory use of certification schemes for tropical timber products, many producer countries agree in principle with the FSC that such schemes need to be global in coverage. However, they go even further and argue that 'global' means not just timber products but also their close substitutes (*INER* 1994 and *MTIDC News* 1994). In other words, whereas the FSC would like to see certification used to ensure sustainable forest management worldwide, many producer countries see value in certification only if consistent environmental criteria are applied to both tropical timber and its competing consumer products. This view is expressed eloquently by Dr Lim Keng Yaik, Malaysia's Minister of Primary Industries:

> Certification is designed to enable consumers to make educated choices between products, based on their environmental values...It is therefore important to certify not only tropical timber but also other types, including temperate and boreal timber, and even substitutes such as plastic, aluminium and steel. The aim is to establish a level playing-field for all these competing products. (*MTIDC* News 1994.)

Because certification is seen to be such a key issue in all discussion of trade policy issues, it has also become the focus of recent ITTC Sessions. At the 16th ITTC Session at Cartagena, Colombia, 16–23 May 1994, the issue was further debated, as the result of a consultants' report on *Certification Schemes for All Timber and Timber Products* and subsequent Working Party. Despite the anticipated differences between producer and consumer countries, there was nonetheless broad agreement that an international certification scheme would need to be adopted sooner or later. Moreover, it was agreed that any scheme should be voluntary, respect national sovereignty, conform to GATT trade measures, and be based on transparent, internationally recognized guidelines that are compatible with the specific conditions affecting each producer (*INER* 1994).

Meanwhile, the year 2000 approaches. What remains to be seen is whether the recent intensification of policy debate will lead to equally rapid progress – whether inside ITTO or in other international fora – towards trade-related incentives for sustainable management of tropical and other production forests of the world.

REFERENCES

CHAPTER 1

Amelung, T and Diehl, M, 1991. *Deforestration of Tropical Rainforests: Economic Causes and Impact on Development*, Kieler Studien 241, Tubingen, Mohr, Germany.

Barbier, E B, 1994. 'The Environmental Effects of the Forestry Sector'. In Organization for Economic Cooperation and Development (OECD), *The Environmental Effects of Trade*, OECD, Paris.

Barbier, E B, Burgess, J C and Folke, C, 1994. *Paradise Lost? The Ecological Economics of Biodiversity*. Earthscan Publications, London.

Barbier, E B, Burgess, J C and Markandya, A, 1991. 'The Economics of Tropical Deforestation', *Ambio* 20(2):55–58.

Dembner, S, 1991. 'Provisional Data from the Forest Resources Assessment 1990 Project', *Unasylva* 42(164):40–44.

Gillis, M, 1990. 'Forest Incentive Policies', Paper prepared for the World Bank Forest Policy Paper, The World Bank, Washington, DC.

Hyde, W F, Newman, D H and Sedjo, R A, 1991. *Forest Economics and Policy Analysis: An Overview*, World Bank Discussion Paper No 134, The World Bank, Washington DC.

International Tropical Timber Organization (ITTO), 1990. *Guidelines for the Sustainable Management of Natural Tropical Forests*, ITTO, Yokohoma.

Pearce, D W, 1990. *An Economic Approach to Saving the Tropical Forests*, LEEC Discussion Paper 90–06, London Environmental Economics Centre, UK.

Poore, D, Burgess, P, Palmer, J, Rietbergen, S and Synnott, T, 1989. *No Timber Without Trees: Sustainability in the Tropical Forest*, Earthscan Publications Ltd, London.

Reid, W V, and Miller, K R, 1989. *Keeping Options Alive: The Scientific Basis for Conserving Biodiversity*, World Resources Institute, Washington, DC.

Repetto, R, 1990. 'Deforestation in the Tropics', *Scientific American*, 262(4): 36–45.

Repetto, R and Gillis, M, 1988. *Public Policy and the Misuse of Forest Resources*, Cambridge University Press, Cambridge.

Vincent, J R and Binkley, C S, 1991. *Forest Based Industrialization: A Dynamic Perspective*, Harvard Institute for International Development (HIID) Development Discussion Paper No 389, Harvard University, Cambridge, Massachusetts.

World Resources Institute (WRI), 1992. *World Resources 1992–93*, Oxford University Press, London.

CHAPTER 2

Arnold, M 1991. 'Forestry Expansion: A Study of Technical, Economic and Ecological Factors', Oxford Forestry Institute Paper No 3, Oxford.

Barbier, E B, Burgess, J C, Aylward, B A, Bishop, J T, and Bann, C 1993. *The Economic Linkages Between the Trade in Tropical Timber and the Sustainable Management of Tropical Forests*, Final Report: ITTO Activity PCM(XI)/4, ITTO, Yokohoma.

Bourke, J, 1992. 'Restrictions on Trade in Tropical Timber', Paper for African Forestry and Wildlife Commission, Rwanda.

Cardellichio, P, Youn, Y, Adams, D, Joo, R and Chmelik, J T, 1989. 'A Preliminary Analysis of Timber and Timber Products Production, Consumption, Trade and Prices in the Pacific Rim until 2000', Working Paper 22, Centre for International Trade in Forest Products, University of Washington, Seattle.

Economic Commission for Europe (ECE)/FAO, 1986. *European Timber Trends and Prospects to the Year 2000 and Beyond*, ECE/TIM/30 United Nations, New York.

ECE/FAO, 1989. *Outlook for the Forest and Forest Products Sector of the USSR*, ECE/TIM/48, United Nations, New York.

ECE/FAO, 1990. *Timber Trends and Prospects for North America*, ECE/TIM/53, United Nations, New York.

FAO, 1988. *An Interim Report on the State of Forest Resources in Developing Countries*, FAO, Rome, Italy.

FAO, 1990. *Forest Products Prices 1969–88*, FAO Forestry Paper 95, FAO, Rome, Italy.

FAO, 1991. *Forest Products: World Outlook Projections: Projections of Consumption and Production of Wood-Based Products to 2010*, FAO Forestry Paper 84, Vol: 1 and 2, FAO, Rome, Italy.

FAO, 1992a. 'The Forest Resources of the Tropical Zone by Main Ecological Regions', Forest Resources Assessment 1990 Project, FAO, Rome, Italy.

FAO. 1992b. *AGROSTAT Computerized Information Series: Forest Products*, FAO, Rome, Italy.

Kallio, M, Dykstra, D and Binkley, C 1987. *The Global Forestry Sector: An Analytical Perspective*, John Wiley and Sons, New York.

Perez-Garcia, J M and Lippke, B R 1993. *Measuring the Impacts of Tropical Timber Supply Constraints, Tropical Timber Trade Constraints and Global Trade Liberalization*, LEEC Discussion Paper 93–03, London Environmental Economics Centre, UK.

Schmidt, R, 1990. 'Sustainable Management of Tropical Moist Forests', presentation for ASEAN Sub-regional Seminar, Indonesia.

Sedjo, R A and Lyon, K S 1990. *The Long-Term Adequacy of World Timber Supply*, Resources for the Future, Washington DC.

USDA Forest Service, 1990. *An Analysis of the Timber Situation in the United States, 1989–2040*, USDA Forest Service General Technical Report RM–199.

Vincent, J R, 1991. *Tropical Timber Trade, Industrialization, and Policies*, draft paper, HIID, Massachusetts.

World Bank, 1992a. *World Tables*. World Bank, Washington DC.

World Bank, 1992b. *World Development Report 1992: Development and the Environment*, World Bank, Washington DC.

(WRI) 1992. *World Resources 1992–3*, Oxford University Press, London.

CHAPTER 3

Allen, J and Barnes, D, 1985. 'The Causes of Deforestation in Developing Countries', *Annals of the Association of American Geographers* 75(2) pp163–84.

Amelung, T and Diehl, M, 1992. *Deforestation of Tropical Rainforests: Economic Causes and Impact on Development*, Tubingen, Mohr.

Barbier, E B, 1991. 'Tropical Deforestation', Chapter 8 in D W Pearce (ed), *Blueprint 2: Greening the World Economy*, Earthscan Publications, London.

Barbier, E B, 1993. 'Economic Aspects of Tropical Deforestation in Southeast Asia', *Global Ecology and Biogeography Letters* 3:1–20.

Barbier, E, 1994. 'The Environmental Effects of the Forestry Sector', in Organization for Economic Cooperation and Development (OECD), *The Environmental Effects of Trade*, OECD, Paris.

Barbier, E B, Burgess, J C, Aylward, B A, Bishop, J T, and Bann, C 1993. *The Economic Linkages Between the Trade in Tropical Timber and the Sustainable Management of Tropical Forests*, Final Report: ITTO Activity PCM(XI)/4, ITTO, Yokohoma.

Barbier, E B, Bockstael, N, Burgess, J C and Strand, I, 1994. The Linkages between the Timber Trade and Tropical Deforestation – Indonesia forthcoming in *World Economy*.

Barbier, E B, Rietbergen, S and Pearce, D W, 1994. 'Economics of Tropical Forest Policy', in J B Thornes (ed), *Deforestation: Environmental and Social Impacts*, Chapman and Hall, London.

Barbier, E B, Burgess, J C and Folke, C, 1994. *Paradise Lost? the Ecological Economics of Biodiversity*, Earthscan Publications, London.

Barbier, E B, Burgess, J C and Markandya, A, 1991. 'The Economics of Tropical Deforestation', *Ambio* 20(2):55–58.

Binswanger, H 1989. *Brazilian Policies that Encourage Deforestation in the Amazon*, World Bank Environment Department Working Paper No 16, Washington DC.

Burgess, J C, 1993. 'Timber Production, Timber Trade and Tropical Deforestation', *Ambio*, Vol 22, No 2–3, pp 136–43.

Burgess, J C, 1994. 'Timber Trade as a Cause of Tropical Deforestation', in C Perrings, K-G Mäler, C Folke, C S Holling and B-O Jansson (eds), *Biodiversity Conservation: Policy Issues and Options*, Kluwer Academic Press, Amsterdam.

Capistrano, A D, 1990. 'Macroeconomic Influences on Tropical Forest Depletion: A Cross Country Analysis', PhD Dissertation, University of Florida, USA.

Capistrano, A D and Kiker, C F, 1990. 'Global Economic Influences on Tropical Closed Broadleaved Forest Depletion', 1967–85, Food Resources Economics Department, University of Florida, USA.

FAO, 1988 *An Interim Report on the State of Forest Resources in Developing Countries*, FAO, Rome.

FAO, 1992a. 'The Forest Resources of Tropical Zone by Main Ecological Regions', Forest Resources Assessment 1990 Project, FAO, Rome, Italy.

FAO, 1992b. *Forests Products Yearbook*, 1990, FAO, Rome.

FAO/UNEP, 1981. *Tropical Forest Resources Assessment Project*, FAO, Rome.

Hyde, W F, Newman, D H and Sedjo, R A, 1991. *Forest Economics and Policy Analysis: An Overview*, World Bank Discussion Paper No 134, World Bank, Washington DC.

Mahar, D, 1989. 'Deforestation in Brazil's Amazon Region: Magnitude, Rate and Causes', in G Schramm and J Warford (eds), *Environmental Management and Economic Development*, John Hopkins University Press, Baltimore, USA.

Palo, M, Mery, G and Salmi, J, 1987. 'Deforestation in the Tropics: Pilot Scenarios Based on Quantitative Analysis', in M Palo and J Salmi (eds), *Deforestation or Development in the Third World*, Division of Social Economic of Forestry, Finnish Forest Research Institute, Helsinki.

Pearce, D W, 1990. 'An Economic Approach to Saving the Tropical Forests', *LEEC Discussion Paper 90–06*, LEEC, UK.

Pearce, D W, Barbier, E B and Markandya, A, 1990. *Sustainable Development: Economics and Environment in the Third World*, Elgar and Earthscan, London.

Poore D, Burgess, P, Palmer, J, Rietbergen, S and Synnott, T, 1989. *No Timber Without Trees: Sustainability in the Tropical Forest*, Earthscan, London.

Reid, W V and Miller, K R, 1989. *Keeping Options Alive: The Scientific Basis for Conserving Biodiversity*, World Resources Institute, Washington DC.

Reis, E J and Marguilis, S, 1991. *Options for Slowing Amazon Jungle Clearing*, paper prepared for conference on Economic Policy Responses to Global Warming, Rome.

Repetto, R, 1990a. 'Macroeconomic Policies and Deforestation', paper prepared for UNU/WIDER Project *The Environment and Emerging Development Issues*.

Repetto, R, 1990b, 'Deforestation in the Tropics', *Scientific American*, 262(4):36–45.

Repetto, R and Gillis, M, 1988. *Public Policies and the Misuse of Forest Resources*, Cambridge University Press, Cambridge.

Schneider, R, McKenna, J, Dejou, C, Butler, J and Barrows, J, 1990. *Brazil: An Economic Analysis of Environmental Problems in the Amazon*, World Bank, Washington DC.

Southgate, D, Sierra, R and Brown, L, 1989. 'The Causes of Tropical Deforestation in Ecuador: A Statistical Analysis', *LEEC Discussion Paper 89–09*, LEEC, UK.

Southgate, D, 1991. 'Tropical Deforestation in Latin America', *LEEC Discussion Paper 90–01*, LEEC, UK.

Vincent, J R, 1990. 'Don't Boycott Tropical Timber', *Journal of Forestry*, 88(4) 56.

Vincent, J R and Binkley, C S, 1991. *Forest Based Industrialization: A Dynamic Perspective*, HIID Development Discussion Paper No. 389, Harvard University, Cambridge, Massachusetts.

World Bank, 1991. *World Tables*, Washington DC.

World Resources Institute, 1990. *World Resources* 1990–91, Oxford University Press, London.

CHAPTER 4

Alexander, S and Greber, B, 1991. 'Environmental Ramifications of Various Materials Used in Construction and Manufacture in the United States', Gen. Tech. Rep. PNW-GTR-277, Portland, US Dept of Agric Forest Service, Pacific Northwest Research Station.

Barbier, E B, Bockstael, N, Burgess, J C and Strand, I, 1994. The Linkages between the Timber Trade and Tropical Deforestation – Indonesia forthcoming in *World Economy*.

Barbier, E B, Burgess, J C, Aylward, B A, Bishop, J T, and Bann, C, 1993b. *The Economic Linkages Between the Trade in Tropical Timber and the Sustainable Management of Tropical Forests*, Final Report: ITTO Activity PCM(XI)/4, ITTO, Yokohoma.

Brooks, D, 1993. *Market Conditions for Tropical Timber*, LEEC Discussion Paper 93–04, LEEC, London.

Buongiorno, J, and Manurung, T, 1992. 'Predicted Effects of an Import Tax in the European Community on the International Trade in Tropical Timbers', *mimeo*, Department of Forestry, University of Wisconsin, Madison.

Buttoud, G and Mamoudou, H, 1986. 'Trade in Forest Products among Developing Countries'. *Unasylva*, 38 (153):20–27.

Cardellichio, P A, Youn, Y C, Adams, D M, Joo, R W and Chmelik, J T, 1989. *A Preliminary Analysis of Timber and Timber Products Production, Consumption, Trade, and Prices in the Pacific Rim Until 2000*, CINTRAFOR Working Paper 22, Center for International Trade in Forest Products, University of Washington, Seattle.

Constantino, L F, 1988. 'Analysis of the International and Domestic Demand for Indonesian Wood Products', *mimeo*, Report for the FAO, Dept of Rural Economy, University of Alberta, Edmonton, Alberta.

Cooper, R J, 1990. 'High Value Markets for Tropical Sawnwood, Plywood, and Veneer in the European Community'. Report for FAO, Forest Industries Division, Rome, Italy.

ECE/FAO. *TIMTRADE Database*, United Nations, Economic Commission for Europe, Geneva.

FAO, 1990. *Forest Products Prices, 1969–1988,* Forestry Paper 95, FAO, Rome, Italy.

FAO, 1991. *Forest Products: World Outlook Projections* (update, September, 1991), Forestry Paper 84, FAO, Rome, Italy.

FAO, 1992. *Yearbook of Forest Products*, FAO, Forestry Series 25, Rome, Italy.

Foreign Agricultural Service, 1990. 'Wood Products: International Trade and Foreign

Markets', Working Paper 490, US Department of Agriculture, Foreign Agriculture Service, Washington DC.

Forestry Canada, 1992. *Selected Forestry Statistics Canada, 1991*, Inf Rep E-X-46, Forestry Canada, Policy and Economics Directorate, Ottawa, Onatario.

Friends of the Earth (FOE), 1992. *Press Release*, 28 March 1992.

Ingram, D 1993. 'Historical Price Trends of Nonconiferous Tropical Logs and Sawn Wood Imported to the United States, Europe, and Japan', General Technical Report, USDA Forest Service, Forest Products Laboratory, Madison, WI.

ITTO, 1990. *Guidelines for the Sustainable Management of Natural Tropical Forests*, ITTO, Yokohoma.

ITTO, 1991. 'Study of the Trade and Markets for Tropical Hardwoods in Europe', ITTO, Yokohoma.

ITTO, 1992. 'Results of the 1991 Forecasting Enquiry for the Annual Review: Tropical Timber Market Worksheets, 1990–1992', ITTC (XII)/4, 22, ITTO, Yokohama.

Milland Fine Timber Ltd, 1990. *New Business Feasibility Study*, prepared by The Business Research Unit, 5 St John's Land, London.

Japan Wood-Products Information and Research Centre (JAWIC), 1991. 'Wood Supply and Demand Information Service'. JAWIC, Japan.

Japan Lumber Reports, 1992. No 142 (2/21/92).

Jen, I, 1988. 'Taiwan Timber Production, Timber Prices, and Forest Products International Trade Statistics, 1972–1987', Taiwan Forestry Research Institute, Taipei.

Nectoux, F, Kuroda, Y, 1989. 'Timber from the South Seas: An Analysis of Japan's Tropical Timber Trade and its Environmental Impact', WWF International, Gland, Switzerland.

NEI, 1989. *An Import Surcharge on the Import of Tropical Timber in the European Community: An Evaluation*, NEI, Rotterdam.

Pringle, S I, 1976. 'Tropical Moist Forests in World Demand, Supply, and Trade', *Unasylva* 28 (112–113): 106–118).

Stichting Bos en Hout (SBH), 1991. 'Market Intelligence: Analysis of the Wood Flow as a Basis for an Early Warning System for the Tropical Timber Market', Final report, ITTO Project PD14/87(m), SBH, Wageningen, Netherlands.

Vincent, J R, 1989. 'Optimal Tariffs on Intermediate and Final Goods: The Case of Tropical Forest Products', *Forest Science*, 35(3):720–731.

Vincent, J R, 1992. 'A Simple, Nonspatial Modelling Approach for Analysing a Country's Forest-Products Trade Policies', in R Haynes, P Harou and J Mikowski (eds), *Forestry Sector Analysis for Developing Countries*, Proceedings of Working Groups, Integrated Land Use and Forest Policy and Forest Sector Analysis Meetings, 10th Forestry World Congress, Paris.

Vincent J, Brooks, D and Gandapur, and Alamgir K, 1991. 'Substitution between Tropical and Temperate Sawlogs', *Forest Science*, 37:1484–1491.

Vincent, J, Gandapur, A K and Brooks, D J 1990. 'Species Substitution and Tropical Log Imports by Japan', *Forest Science* 36(3):657–664.

Ward Associates, J V, 1990. 'The Japanese Market for Tropical Timber: An Assessment for the International Tropical Timber Organization', Washington DC.

Ward International, 1992. 'The North American Market for Tropical Timber: An Assessment for ITTO', Washington DC.

Wood-Products Stockpile Corporation (WSC), 1991. 'International Trade in Wood Products', WSC, New York.

World Wide Fund for Nature (WWF), 1990. *Tropical Rain Forests and the Environment: A Survey of Public Attitudes*, WWF, Godalming, UK.

Youn, Y C and Yum, S C, 1992, 'A Study on the Demand and Supply of Timber in South Korea', paper presented at the symposium 'Forest Sector, Trade and Environmental Impact Models: Theory and Applications', Centre for International Trade in Forest Products, University of Washington, Seattle, 30 April – 1 May.

CHAPTER 6

Arnold, M, 1990. *Common Property Management and Sustainable Development in India*, Forestry for Sustainable Development Program Working Paper 9, University of Minnesota.

Asia-Pacific Action Group, 1990. *The Barnett Report: A Summary of the Commission of Inquiry into Aspects of the Timber Industry in Papua New Guinea*, Asia-Pacific Action Group, Hobart, Australia.

Aylward, B A, Bishop, J C, and Barbier, E B, 1993. *Economic Efficiency, Rent Capture and Market Failure in Tropical Forest Management*. LEEC Gatekeeper Series 93–01, London.

Barbier, E B, 1992. 'Tropical Rain Forest Utilization', in T M Swanson and E B Barbier (eds), *Economics for the Wilds: Wildife, Wildlands, Diversity and Development*, Earthscan Publications, London.

Barbier, E B, 1993. 'Economic Aspects of Tropical Deforestation in Southeast Asia', *Global Ecology and Biogeography Letters* 3:1–20.

Barbier, E B, 1994. 'The Economics of Environment and Development', in A-M Jansson, M Hammer, C Folke and R Costanza, *Investing in Natural Capital: The Ecological Economics of Sustainability*, Island Press, Washington DC.

Barbier, E B, Burgess, J C and Markandya, A, 1991. 'The Economics of Tropical Deforestation', *AMBIO* 20(2):55–58.

Bhat, M G and Huffaker, R G, 1991. 'Private Property Rights and Forest Preservation in Karnataka Western Ghats, India', *American Journal of Agricultural Economics*, May: 375–387.

Bishop, J, Aylward, B and Barbier, E B, 1991. *Guidelines for Applying Environmental Economics in Developing Countries*, LEEC Gatekeeper Series 91–02, London.

Cardellichio, P A, Youn, Y C, Adams, D A, Joo, R W and Chmelik, J T, 1989. *A Preliminary Analysis of Timber and Timber Products Production, Consumption, Trade, and Prices in the Pacific Rim Until 2000*, CINTRAFOR, Washington DC.

Capistrano, D A and Kiker, C F, 1990. *Global Economic Influences on Tropical Closed Broad Leaved Forest Depletion, 1967–1985*, University of Florida.

Cruz, W and Repetto, R, 1992. *The Environmental Effects of Stabilization and Structural Adjustment Programs: The Philippines Case*, WRI, Washington DC.

Gillis, M, 1988. 'Indonesia: Public Policies, Resource Management, and the Tropical Forest', in R Repetto and M Gillis (eds), *Public Policies and the Misuse of Forest Resources*, Cambridge University Press, Cambridge.

Gillis, M, 1990. 'Forest Incentive Policies', Paper prepared for the World Bank Forest Policy Paper, World Bank, Washington DC.

Grut, M, 1990. *Economics of Managing the African Rainforest*, Paper prepared for the Thirteenth Commonwealth Forestry Conference and Revised in February 1990.

Grut, M, Gray, J A and Egli, N, 1991. *Forest Pricing and Concession Policies: Managing the High Forests of West and Central Africa*, World Bank Technical Paper Number 143, Africa Technical Department Series, World Bank, Washington DC.

Hyde, W F, Newman, D H and Sedjo, R A, 1991. *Forest Economics and Policy Analysis: An Overview*, World Bank Discussion Paper 134, World Bank, Washington DC.

ITTO, 1990. *Guidelines for the Sustainable Management of Natural Tropical Forests*, ITTO, Yokohama.

Jonish, J, 1992. *Sustainable Development and Employment: Forestry in Malaysia*, World Employment Programme Working Paper, International Labour Office, Geneva.

Kahn, J and McDonald, J, 1990. *Third World Debt and Tropical Deforestation*, Department of Economics, SUNY – Binghamton, NY.

Paris, R and Ruzicka, I, 1991. *Barking Up the Wrong Tree: The Role of Rent Appropriation in Sustainable Forest Management*, Environment Office Occasional Paper No 1, Asian Development Bank, Manila, Philippines.

Pearce, D W, 1990. *An Economic Approach to Saving the Tropical Forests*, LEEC Gatekeeper 90–06, LEEC, London.

Poore, D, Burgess, P, Palmer, J, Rietbergen, S and Synnott, T, 1989. *No Timber Without Trees: Sustainability in the Tropical Forest*, Earthscan, London.

Repetto, R, 1990. 'Deforestation in the Tropics', *Scientific American*, 262(4):36–45.

Repetto, R and Gillis, M, 1988. *Public Policies and the Misuse of Forest Resources*, Cambridge University Press, Cambridge.

Vincent, J R and Binkley, C S, 1991. *Forest-based Industrialization: A Dynamic Perspective*, HIID Development Discussion Paper No 389, Harvard University, Cambridge, MA.

WRI and TSC, 1991. *Accounts Overdue: Natural Resource Depreciation in Costa Rica*, WRI, Washington.

Yao, J, Kouadio, Y, Akomian, J and Angoran, O, 1991. *Structural Adjustment and the Environment: The Case of Côte d'Ivoire*, consultant report to WWF, and LEEC, Abidjan, and London.

CHAPTER 7

Anderson, K, 1992. 'The Standard Welfare Economics of Policies Affecting Trade and the Environment', Chapter 2 in K Anderson and R Blackhurst (eds), *The Greening of World Trade Issues*, Harvester Wheatsheaf, London.

Barbier, E B 1987. 'Natural Resources Policy and Economic Framework', Annex 1 in J Tarrant et al, *Natural Resources and Environmental Management in Indonesia*, USAID, Jakarta.

Barbier, E B, 1993.'Economic Aspects of Tropical Deforestation in Southeast Asia', *Global Ecology and Biogeography Letters* 3:1–20.

Barbier, E, 1994. 'The Environmental Effects of the Forestry Sector', in *The Environmental Effects of Trade*, OECD, Paris.

Barbier, E B, Bockstael, N, Burgess, J C and Strand, I, 1994. The Linkages between the Timber Trade and Tropical Deforestation – Indonesia forthcoming in *World Economy*.

Barbier, E B, Burgess, J C, Aylward, B A, Bishop, J T, and Bann, C, 1993b. *The Economic Linkages Between the Trade in Tropical Timber and the Sustainable Management of Tropical Forests*, Final Report: ITTO Activity PCM(XI)/4, ITTO, Yokohoma.

Bourke, I J, 1988. *Trade in Forest Products: A Study of the Barriers Faced by the Developing Countries*, FAO Forestry Paper 83, FAO, Rome.

Boyd, R, Hyde, W F and Krutilla, K, 1991. 'Trade Policy and Environmental Accounting: A Case Study of Structural Adjustment and Deforestation in the Philippines', Draft Paper, Economics Dept, Ohio University, Athens, Ohio.

Burgess, P F, 1989. 'Asia', Chapter 5 in D Poore et al, *No Timber Without Trees: Sustainability in the Tropical Forest*, Earthscan, London.

Constantino, L F, 1988a. 'Analysis of the International and Domestic Demand for Indonesian Wood Products', *mimeo*, Report for the FAO, Dept of Rural Economy, University of Alberta, Edmonton, Alberta.

Constantino, L F, 1988b. 'Demand, Supply and Development Issues in the Indonesian Forest Sector', *mimeo*, Report for the FAO, Dept of Rural Economy, University of Alberta, Edmonton, Alberta.

Constantino, L F, 1990. *On the Efficiency of Indonesia's Sawmilling and Plymilling Industries*, D G of Forest Utilization, Ministry of Forestry, Government of Indonesia and FAO, Jakarta.

Constantino, L F and Ingram, D, 1990. *Supply-Demand Projections for the Indonesian Forest Sector*, D G of Forest Utilization, Ministry of Forestry, Government of Indonesia and FAO, Jakarta.

Fitzgerald, B, 1986. *An Analysis of Indonesian Trade Policies: Countertrade, Downstream Processing, Import Restrictions and the Deletion Program*, CPD Discussion Paper 1986–22, World Bank, Washington DC.

Gillis, M, 1988. 'Indonesia: Public Policies, Resource Management, and the Tropical Forest', in R Repetto and M Gillis (eds), *Public Policies and the Misuse of Forest Resources*, Cambridge University Press, Cambridge.

Gillis, M, 1990. 'Forest Incentive Policies', Paper prepared for the World Bank Forest Policy Paper, World Bank, Washington DC.

Gray, J A and Hadi, S, 1990. *Fiscal Policies and Pricing in Indonesian Forestry*, D G of Forest Utilization, Ministry of Forestry, Government of Indonesia and FAO, Jakarta.

Grut, M, Gray, J A and Egli, N, 1991. *Forest Pricing and Concession Policies: Managing the High Forests of West and Central Africa*, World Bank Technical Paper No 143, Africa Technical Department Series, World Bank, Washington DC.

ITTO, 1990. *Guidelines for the Sustainable Management of Natural Tropical Forests*, ITTO, Yokohama.

Perez-Garcia, J M and Lippke, B R, 1993. *Measuring the Impacts of Tropical Timber Supply Constraints, Tropical Timber Trade Constraints and Global Trade Liberalization*, LEEC Discussion Paper 93–03, LEEC, London.

Repetto, R and Gillis, M, 1988. *Public Policies and the Misuse of Forest Resources*, Cambridge University Press, Cambridge.

Ruitenbeek, H J, 1989. 'Social Cost-Benefit Analysis of the Korup Project, Cameroon', prepared for WWF and Republic of Cameroon, London.

Sedjo, R A, 1987. 'Incentives and Distortions in Indonesian Forest Policy', *mimeo*, Resources for the Future, Washington DC.

Vincent, J R, 1989. 'Optimal Tariffs on Intermediate and Final Goods: The Case of Tropical Forest Products', *Forest Science* 35(3):720–731.

Vincent, J R, 1992a. 'A Simple, Nonspatial Modelling Approach for Analysing a Country's Forest-Products Trade Policies', in R Haynes, P Harou and J Mikowski (eds), *Forestry Sector Analysis for Developing Countries*, Proceedings of Working Groups, Integrated Land Use and Forest Policy and Forest Sector Analysis Meetings, 10th Forestry World Congress, Paris.

Vincent, J R, 1992b. 'The Tropical Timber Trade and Sustainable Development', *Science*, 256:1651–1655.

Vincent, J R and Binkley, C S, 1991. *Forest Based Industrialization: A Dynamic Perspective*, Development Discussion Paper No 389, HIID, Cambridge, Massachusetts.

Vincent, J R, Brooks, B J and Gandapur, A K, 1991. 'Substitution Between Tropical and Temperate Sawlogs' *Forest Science* 37(5):1484–1491.

Vincent, J R, Gandapur, A K and Brooks, D J, 1990. 'Species Substitution and Tropical Log Imports by Japan', *Forest Science* 36(3):657–664.

World Bank, 1989. *Indonesia – Forest Land and Water: Issues in Sustainable Development*, World Bank, Washington DC.

CHAPTER 8

Barbier, E B, Burgess, J C, Aylward, B A, Bishop, J T and Bann, C, 1993. *The Economic Linkages Between the Trade in Tropical Timber and the Sustainable Management of Tropical Forests*, Final Report: ITTO Activity PCM(XI)/4, ITTO, Yokohoma.

Bourke, I J, 1988. *Trade in forest products: a study of the barriers faced by the developing countries*, FAO Forestry Paper 83, FAO, Rome.

Bourke, I J, 1992a. *Comments on the Current Situation Regarding Trade Barriers Facing Forest Products*, unpub mimeo, FAO, Rome.

Bourke, I J, 1992b. *Restriction on Trade in Tropical Timber*, Secretariat Note, African Forestry and Wildlife Commission, 9th Session, Kigali, Rwanda, 10–14 August.

Constantino, L F, 1988. *Analysis of the International and Domestic Demand for Indonesian Wood Products*, unpub mimeo, FAO Project INS/83/019.

Flora, D F and McGinnis, W.J, 1991. *Effects of Spotted-Owl Reservations, The State Log Embargo, Forest Replanning and Recession on Timber Flows and Prices in the Pacific Northwest and Abroad*, unpub review draft, Trade Research, Pacific Northwest Research Station, USDA Forest Service, Seattle.

NEI, 1989. *An Import Surcharge on the Import of Tropical Timber in the European Community: An Evaluation*, Netherlands Economic Institute (NEI) unpub mimeo, Rotterdam.

Perez-Garcia, J M, 1991. *An Assessment of the Impacts of Recent Environmental and Trade Restrictions on Timber Harvests and Exports*, Working Paper No 33, Center for International Trade in Forest Products, University of Washington, Seattle.

Perez-Garcia J M and Lippke, B R, 1993. *Measuring the Impacts of Tropical Timber Supply Constraints, Tropical Timber Trade Constraints and Global Trade Liberalization*, LEEC Discussion Paper 93-03, LEEC, London.

Vincent, J R and Binkley, C S, 1991. *Forest-Based Industrialization: A Dynamic Perspective*, Development Discussion Paper No 389, HIID, Cambridge.

CHAPTER 9

Arden-Clarke, C, 1991. *The General Agreement on Tariffs and Trade, Environmental Protection and Sustainable Development*, WWF Discussion Paper, WWF International, Geneva.

Barbier, E, 1994. 'The Environmental Effects of the Forestry Sector', in *The Environmental Effects of Trade*, OECD, Paris.

Barbier, E B, Bockstael, N, Burgess, J C and Strand, I, 1994. The Linkages between the Timber Trade and Tropical Deforestation – Indonesia, forthcoming in *World Economy*.

Barbier, E B, Burgess, J C, Aylward, B A, Bishop, J T, and Bann, C, 1993. *The Economic Linkages Between the Trade in Tropical Timber and the Sustainable Management of Tropical Forests*, Final Report: ITTO Activity PCM(XI)/4, ITTO, Yokohoma.

Barbier, E B and Rauscher, M, 1994. 'Trade, Tropical Deforestation and Policy Interventions'. *Environmental and Resource Economics* 4:75–90.

Dudley, N, 1991. *Importing Deforestation: Should Britain Ban the Import of Tropical Hardwoods?*, Study for WWF UK, Godalming.

Flora, D F and McGinnis, W J, 1991. 'Effects of Spotted-Owl Reservations, The State-Log Embargo, Forest Replanning, and Recession on Timber Flows and Prices in the Pacific

Northwest and Abroad', Review Draft, Trade Research, Pacific Northwest Research Station, USDA Forest Service, Seattle.

Government of the Netherlands, 1991. *Policy Paper on Tropical Rain Forests: Summary*, Government of the Netherlands, The Hague.

Grimmett, J J, 1991. *Environmental Regulation and the GATT*, CRS Report for Congress, Congressional Research Service, Library of Congress, Washington DC.

Hewett, J, Rietbergen, S and Baldock, D, 1991. *Target 1995: Problems and Perspectives Relating to the Import of Sustainably Produced Tropical Timber in the Netherlands*, IEEP and IIED, London.

ITTO, 1990. *Guidelines for the Sustainable Management of Natural Tropical Forests*, ITTO, Yokohama.

Perez-Garcia, J M, 1991. *An Assessment of the Impacts of Recent Environmental and Trade Restrictions on Timber Harvests and Exports*, Working Paper 33, CINTRAFOR, University of Washington, Seattle.

Perez-Garcia, J M and Lippke, B R, 1993. *Measuring the Impacts of Tropical Timber Supply Constraints, Tropical Timber Trade Constraints and Global Trade Liberalization*, LEEC Discussion Paper 93–03, LEEC, London.

Poore, D, Burgess, P, Palmer, J, Rietbergen, S and Synnott, T, 1989. *No Timber Without Trees: Sustainability in the Tropical Forest*, Earthscan, London.

Wibe, S and Jones, T, 1992. *Forests: Market and Intervention Failures – Five Case Studies*, Earthscan, London.

CHAPTER 10

Amelung, T, 1993. 'Tropical Deforestation as an International Economic Problem'. In H Giersch (ed), *Economic Evolution and Environmental Concerns*, Kiel Institute of World Economics and Kluwer Press Kiel, and Dordrecht.

Barbier, E B, 1993. 'Economic Aspects of Tropical Deforestation in Southeast Asia', *Global Ecology and Biogeography Letters* 3:1–20.

Barbier, E, 1994. 'The Environmental Effects of the Forestry Sector', in *The Environmental Effects of Trade*, OECD, Paris.

Barbier, E B, Burgess, J C, Swanson, T M and Pearce, D W, 1990. *Elephants, Economics and Ivory*, Earthscan Publications, London.

Barbier, E B, Bockstael, N, Burgess, J C and Strand, I, 1994. The Linkages between the Timber Trade and Tropical Deforestation – Indonesia forthcoming in *World Economy*.

Barbier, E B, Burgess, J C, Aylward, B A, Bishop, J T, and Bann, C, 1993. *The Economic Linkages Between the Trade in Tropical Timber and the Sustainable Management of Tropical Forests*, Final Report: ITTO Activity PCM(XI)/4, ITTO, Yokohama.

Barrett, S, 1990. 'The Problem of Global Environmental Protection', *Oxford Review of Economic Policy* 6(1):68–79.

Buongiorno, J and Manurung, T, 1992. 'Predicted Effects of an Import Tax in the European

Community on the International Trade of Tropical Timbers', *mimeo*, Department of Forestry, University of Wisconsin, Madison.

Burgess, J C, 1992. 'The Environmental Effects of Trade in Endangered Species' In OECD *The Environmental Effects of Trade*, OECD, Paris

Constantino, L F, 1990. *On the Efficiency of Indonesia's Sawmilling and Plymilling Industries*, D G of Forest Utilization, Ministry of Forestry, Government of Indonesia and FAO, Jakarta.

FAO, 1992. *Forest Products Yearbook 1990*. FAO, Rome.

Ferguson, I S and Muñoz-Reyes Navarro, J, 1992. 'Working Document for ITTO Expert Panel on Resources Needed by Producer Countries to Achieve Sustainable Management by the Year 2000', ITTO, Yokohama.

Gewerkschaft Holz und Kunststoff (GHK), Central Federation of German Timber and Plastics Processing Industries (HDH) and German Timber Importers Federation (VDH) 1992. *Joint Declaration on the Protection of Tropical Forests*, GHK/HDH/VDH, Germany.

ITTC, 1992a. *Report From the Expert Panel on Estimation of Resources Needed by Producer Countries to Attain Sustainability by the Year 2000*, ITTC(XII)/7 Rev 1, 10th Session, 6–14 May, Yaoundé, Cameroon.

ITTC, 1992b. *Report on the 8th Committee of Parties to the Convention on International Trade in Endangered Species, Kyoto, 1–13 March 1992*, PCM(X)/7, 10th Session, 6–14 May, Yaoundé, Cameroon.

ITTC, 1992c. *Report on Preparations for the 1992 United Nations Conference on Environment and Development*, ITTC(XII)/8, 10th Session, 6–14 May, Yaoundé, Cameroon.

ITTC, 1992d. *Report Submitted to the International Tropical Timber Council on the 'Workshop on Issues Related to Incentives to Promote Sustainable Development of Tropical Forests through Trade'*, PCM(X)/4, 10th Session, 6–14 May, Yaoundé, Cameroon.

NEI, 1989. *An Import Surcharge on the Import of Tropical Timber in the European Community: An Evaluation*, NEI, Rotterdam.

OFI in association with the Timber Research and Development Association (TRADA), 1991. *Incentives in Producer and Consumer Countries to Promote Sustainable Development of Tropical Forests*, ITTO Pre-Project Report, PCM, PCF, PCI(IV)/1/Rev 3, OFI, Oxford.

Perez-Garcia, J M and Lippke, B R, 1993. *Measuring the Impacts of Tropical Timber Supply Constraints, Tropical Timber Trade Constraints and Global Trade Liberalization*, LEEC Discussion Paper 93–03, LEEC, London.

Poore, D, Burgess, P, Palmer, J, Rietbergen, S, and Synnott, T, 1989. *No Timber Without Trees: Sustainability in the Tropical Forest*, Earthscan, London.

Sedjo, R A, Bowes, M and Wiseman, C, 1991. *Toward a Worldwide System of Tradeable Forest Protection and Management Obligations*, ENR 91–16, Energy and Natural Resources Division, Resources for the Future, Washington DC.

Vincent, J R, 1990. 'Don't Boycott Tropical Timber', *Journal of Forestry*, 88(4):56.

EPILOGUE

FSC, 1993. *Draft Principles and Criteria of Natural Forest Management*, Fauna & Flora Preservation Society, London.

International Environmental Reporter (INER), 1994. 'Tropical Forests – Producer, Consumer Nations in Conflict over Certification to Achieve Sustainability', *Current Reports* 17(13):551–553, 29 June.

MTIDC News, 1994. 'Certification Efforts Dead Wood Unless Fair Global Approach Adopted,' June.

Synnott, T J, 1993. *Draft Principles and Criteria for Plantation Forests*, Fauna & Flora Preservation Society, London.

Index

administration, forest pricing 67–72
bans, trade *see* interventions, trade
biomass reduction 27&n

cartelization 123–4
case studies
   Congo: CIB 78–81
   Indonesia: total demand effect 50–2
Centre for International Trade in Forest Products (CINTRAFOR) 18&n–19
certification 116–17
   concession certification 151–3
   country 153–5
   product labelling 151
   summary 155
CINTRAFOR Global Trade Model (CGTM) 18n, 130–1, 141
compensation, international 140–50
concessions 65–7
   certification 151–3
   Congo: CIB 78–81
Convention on International Trade in Endangered Species (CITES) 136–7

deforestation, timber trade and 7, 19–20, 21–33
   environmental impacts 24–8
      direct 24–6
      indirect 26–8
   forest management 22–4
      economics of decision making 23–4
      land use options 22–3
   links between 3, 29–32, 74
   conclusions 32–3
developing countries, trade 39–42
domestic market, policy 61–81
   concessions/property rights 65–7
   environmental policy 90–4
   and forest management 61–5
   forest pricing/administration 67–72
   macro-economic policy 64, 72–6
   conclusion 76–7

Annex: Congo - case study 78–81

environment
   domestic policy 90–4
   impacts of timber production 24–8
      direct 24–6
      indirect 2, 26–8
   importing countries' regulations 108–13
   tropical forest policies 58–60
   unilateral interventions 118–25 *passim*
exporting countries, trade intervention in 59, 82–96
   impacts of 86–9
   restrictions 82–6
   trade liberalization 89–94
      domestic environmental policy 90–4
   conclusion 95–6
external funding 148–50

Food and Agricultural Organization of the UN (FAO) 5, 8n, 27n
forest degradation 27&n
forest management 22–4, 59
   domestic market 61–5
   land use options 22–3
   economics of decision making 23–4
forest pricing/administration 67–72
Forest Principles statement, UNCED 149
Forest Stewardship Council (FSC) 161
Friends of the Earth (FOE) 56

General Agreement on Tariffs and Trade (GATT) 98, 115–21, 126, 130, 131, 162
   Tokyo Round 100
   Uraguay Round 161
Generalized System of Preferences (GSP), UNCTAD 98

high-grading 65&n

importing countries, trade intervention 97–113

environmental regulations 108–11
restrictions 97–104
  impacts of 104–6
  non-tariff barriers 101–2
  tariff barriers 98–101
  trends in 102–4
trade liberalization 106–8
conclusions 111–13
Indonesia: total demand effect 50–2
industrialized countries, trade 34–9
International Tropical Timber Organization (ITTO) 3, 37, 122, 132, 136–7, 139, 140, 162
  Target 2000 programme 57, 118, 125–6, 132, 142, 151, 153–4, 158
interventions, trade 115, 132–6, 137–8, 158
  in exporting countries 59, 82–96
    impacts of 86–9
    restrictions 82–6
    trade liberalization 89–94
    conclusion 95–6
  in importing countries 97–113
    environmental regulations 108–11
    impacts of restrictions 104–6
    restrictions 97–104
    trade liberalization 106–8
    conclusions 111–13
  multilateral 125–6, 127
  unilateral 118–25, 126

Korup project, Cameroon 91–2

land use options 22–3
  economics of decision making 23–4
LEEC survey 56–7
liberalization, trade 115, 158
  by exporting countries 89–94
  by importing countries 106–8
  policy options 130–2
  trade bans 132–6

macro-economic policy 64, 72–6
market for timber products 34–57
  in developing countries 39–42
  domestic 61–81
    concessions/property rights 65–7
    environmental policy 90–4
    forest management 61–5
    forest pricing/administration 67–72
    macro-economic policy 72–6
    conclusion 76–7
    Annex: Congo - case study 78–81
  in industrialized countries 34–9
  prices/substitution/demand 42–53

price elasticity 42–6
substitution: by origin 49–50
substitution: non-wood products 48
substitution: other wood products 46–8
total demand effect 50–2
willingness to pay 52–3
conclusions 53–4

Netherlands Economic Institute (NEI) 48
non-tariff barriers 101–2
non-wood products 48, 54

policies, trade
  assessment of options 128–60
    certification 151–5
    'do nothing' scenario 114, 128–9
    quantitative restrictions 115, 136–7
    raising revenues 140–50
    trade bans 132–6
    trade liberalization 130–2
    trade subsidies 138–40
    trade taxes 137–8
    conclusions 155–60
  domestic 61–81
    concessions/property rights 65–7
    environmental policy 90–4
    and forest management 61–5
    forest pricing/administration 67–72
    macro-economic policy 64, 72–6
    conclusion 76–7
    Annex: Congo - case study 78–81
  and the environment 58–60
  conclusions/implications 19–20
prices, consumer 42–53
  forest pricing/administration 67–72
  price elasticity 42–6
  substitution: by origin 49–50
  substitution: non-wood products 48, 54
  substitution: other wood products 46–8
  total demand effect 50–2
  willingness to pay 52–3, 55–7
product labelling 151
property rights 65–7

real price trends 14–15
'rent capture' 123&n
resources and trade 5–20
  changes in 7–8
  future trends 16–19
  production and trade 8–16
    real price trends 14–15
    trade 8–14
    value of trade 15–16
  status of resources 5–7

conclusions/policy implications 19–20
revenues, raising 140–50
    appropriating additional revenue 146–8
    appropriating existing revenues 142–6
    external funding 148–50
royalty 65–7

secondary producers 35n
Spotted owl reservations (US) 110
subsidies, trade 115–16, 138–40, 158
substitution 46–50
    by origin 49–50
    non-wood products 48, 54
    other wood products 46–8
sustainable use of forests 114–27, 131, 149
    identifying/assessing options 114–18, 155–60
        altering trade pattern 114–16
        certification 116–17
        criteria for 117–18
        'do nothing' scenario 28–9, 114
        raising revenue 116
    and multilateral interventions 125–6
    and unilateral interventions 118–25
    willingness to pay 52–3, 55–7
    conclusions 126–7

tariff barriers 98–101
taxes, trade *see* interventions, trade
total demand effect: Indonesia 50–2

trade interventions *see* interventions, trade
transfer pricing 71&n
trespass 65&n
tropical timber trade
    introduction 1–4
    and deforestation 7, 19–20, 21–33
    domestic market 61–81
    export trade interventions 59, 82–96
    import trade interventions 97–113
    market for 34–57
    policies 58–60, 61–81, 128–60
        assessing options 128–60
        domestic 61–81
        the environment 58–60
    resources 5–20
    sustainable use of forests 114–27
    epilogue 161–2

UK Forest Forever Campaign 57
'uncaptured genetic value' 91–2
Union pour le Commerce de Bois Tropicaux (UCBT) 50&n, 147
United Nations
    FAO 5, 8n, 27n
    Forest Principles statement, UNCED 149
    GSP, UNCTAD 98

willingness to pay 52–3, 55–7
World Trade Organization (WTO) 120
World Wide Fund for Nature (WWF) 56

0774